SAFETY OR PROFIT?
International Studies in Governance, Change, and the Work Environment

Edited by
Theo Nichols and David Walters
Cardiff Work Environment Research Centre
Cardiff University

Work, Health, and Environment Series
Series Editors: Charles Levenstein, Robert Forrant,
and John Wooding

Baywood Publishing Company, Inc.
AMITYVILLE, NEW YORK

Copyright © 2013 by Baywood Publishing Company, Inc., Amityville, New York

All rights reserved. No part of this book may be reproduced or utilized in any form or by any means, electronic or mechanical, including photo-copying, recording, or by any information storage or retrieval system, without permission in writing from the publisher. Printed in the United States of America on acid-free recycled paper.

Baywood Publishing Company, Inc.
26 Austin Avenue
PO Box 337
Amityville, NY 11701
(800) 638-7819
E-mail: baywood@baywood.com
Web site: baywood.com

Library of Congress Catalog Number: 2013027439
ISBN 978-0-89503-817-3 (cloth)
ISBN 978-0-89503-818-0 (paper)
ISBN 978-0-89503-819-7 (e-pub)
ISBN 978-0-89503-820-3 (e-pdf)
http://dx.doi.org/10.2190/SOP

Library of Congress Cataloging-in-Publication Data

Safety or profit? : international studies in governance, change, and the work environment / edited by Theo Nichols and David Walters.
        pages cm. -- (Work, health, and environment series)
  Includes bibliographical references and index.
  ISBN 978-0-90503-817-3 (cloth : alk. paper) -- ISBN 978-0-89503-818-0 (pbk.)
   -- ISBN 978-0-89503-819-7 (e-pub) -- ISBN 978-0-89503-820-3 (e-pdf) 1.
Industrial safety. 2. Industrial hygiene. I. Nichols, Theo. II. Walters,
David, 1950-
  T55.S21537   2013
  363.11--dc23
                                        2013027439

Table of Contents

List of Tables and Charts v
Abbreviations vii
Preface ix
Acknowledgments xi
Introduction 1
David Walters and Theo Nichols

PART I. ECONOMIC RESTRUCTURING, LABOR MARKET STRATIFICATION, AND THEIR CONSEQUENCES FOR HEALTH AND SAFETY 15

Chapter 1. Precarity and Workplace Well-Being: A General Review 17
Michael Quinlan

Chapter 2. A Gender Perspective on Work, Regulation, and
Their Effects on Women's Health, Safety and Well-Being. 33
Katherine Lippel and Karen Messing

PART II. NEW GOVERNANCE, ORGANIZED LABOR, DEREGULATION, DECRIMINALIZATION, AND THE NEO-LIBERAL AGENDA 49

Chapter 3. Resilience Within a Weaker Work Environment System—
The Position and Influence of Swedish Safety Representatives 51
Kaj Frick

Chapter 4. Old Lessons for New Governance: Safety or Profit and the
New Conventional Wisdom. 71
Eric Tucker

Chapter 5. Safety, Profits, and the New Politics of Regulation 97
Steve Tombs and David Whyte

Chapter 6. Decriminalization of Health and Safety at Work in Australia . . . 113
Richard Johnstone

iv / SAFETY OR PROFIT?

PART III. THE ROLE AND LIMITS OF EVIDENCE 135

Chapter 7. Competing Interests at Play? The Struggle for Occupational
Cancer Prevention in the UK 137
Andrew Watterson

Chapter 8. The Limits and Possibilities of the Structures and
Procedures for Health and Safety Regulation in
Ontario, Canada 157
Wayne Lewchuk

Chapter 9. From *Piper Alpha* to *Deepwater Horizon* 181
Charles Woolfson

Afterword 205
Theo Nichols and David Walters

References 217
Contributors 249
Index 253

List of Tables and Charts

TABLES

TABLE 1.1 - Key Elements of the PDR Model 30

TABLE 4.1 - Ideal Types of OHS Regimes 73

TABLE 6.1 - Total Number of Workplace Inspections
Made by Work Health and Safety Inspectors
in 2008–2009, Australian States 116

TABLE 6.2 - Enforcement Action Taken by Work Health and Safety
Inspectorates in Australia 2001–2010 by Jurisdiction 118

TABLE 7.1 - Estimates of the Proportion of Cancer Attributable to Occupation . 142

TABLE 7.2 - Competing Interests that Affect Cancer Prevention Activities . . 144

CHARTS

CHART 4.1 - Ontario Ministry of Labour, OHS Inspections and
Orders Issued, 1988/89—2009/10 87

CHART 4.2 - Ontario Ministry of Labour, OHS Stop-Work
Orders Issued, 1988/89—2009/10 87

CHART 4.3 - Ontario Ministry of Labour, OHS Convictions and
Annual Fines Issued, 1993—2008/9 88

CHART 4.4 - Ontario Ministry of Labour, OHS Convictions, Average
Fine per Conviction, 1997—2008/9 88

CHART 4.5 - Ontario Ministry of Labour, Work Refusals Reported, 1984/85—2009/10 91

CHART 4.6 - Annual Average Hours Lost to Labour Disputes, Canada, 1976—2010 (per employed worker). 92

CHART 7.1 - Global Estimated Work-Related Fatality by Cause 140

CHART 7.2 - Number of Cancer Deaths, and Number and Fraction Attributed to Work by Region and Gender 142

CHART 7.3 - Weighted Filtration of Data 155

Abbreviations

ACT	Australian Capital Territory
API	American Petroleum Institute
ASTMS	Association of Scientific, Technical and Managerial Staffs
ATK	Arbetstagarkonsult
AWL	Area-wide leasing
BandCRAO	Building and Concrete Restoration Association of Ontario
BCOHS	Business Council on Occupational Health and Safety
BOEM	Bureau of Ocean Energy Management
BOEMRE	Bureau for Ocean Energy Management, Regulation and Enforcement
BOP	Blowout preventer
BSEE	Bureau of Safety and Environmental Enforcement
CAA	Civil Aviation Authority
CBI	Confederation of British Industry
COPD	Chronic Obstructive Pulmonary Disease
CSR	Corporate social responsibility
DCS	Demand Control Support
DWP	Department of Work & Pensions
EEF	Engineering Employers Federation
EDS	Emergency disconnect system
EMCONET	Employment Conditions Network
ERI	Effort Reward Imbalance
ERS	External responsibility system
EU	European Union
FOD	Field Operations Directorate
FSB	Federation of Small Businesses
GDP	Gross Domestic Profit
GMB	General Municipal and Boilermakers Union
HMFI	Her Majesty's Factory Inspectorate
HRM	Human Resource Management
HSAs	Health and Safety Associations
HSC	Health & Safety Commission
HSE	Health & Safety Executive
HSRs	Health and safety representatives
HSW Act	Health and Safety at Work Act
IADC	International Association of Drilling Contractors
IALL	International Association of Labour Legislation
IARC	International Agency for Research on Cancer

ILO	International Labour Organisation
IRS	Internal responsibility system
JHSC	Joint health and safety committees
LO	Swedish Trade Union Confederation
MOL	Ministry of Labour
MMS	Minerals Management Service
NEER	New Experimental Experience Rating (NEER) Program
NEPA	National Environmental Policy Act
NHS	National Health Service
NGOs	Non-governmental organizations
OCC	Ontario Chamber of Commerce
OECD	Organisation for Economic Co-operation and Development
OH	Occupational Health
OHS	Occupational Health & Safety
OLRB	Ontario Labour Relations Board
ONA	Ontario Nurses' Association
OPEC	Organization of the Petroleum Exporting Countries
OPSEU	Ontario Public Service Employees Union
ORR	Office of Rail Regulation
OSHA	Occupational Safety and Health Administration
PDR model	Pressure, disorganisation and regulatory failure model
PPE	Personal protective equipment
R&D	Research & development
RIAs	Regulatory Impact Assessments
RIDDOR	Reporting of Injuries, Diseases and Dangerous Occurrences Regulations
RSRs	Regional safety representatives
SACO	Swedish Confederation of Professional Associations
SAE	Suspended access equipment
SEMP	Safety and environmental management program
SEMS	Safety and Environmental Management System
SOHAS	Sheffield Occupational Health Project
SRs	Safety representatives
SRSC	Safety Representative and Safety Committee
SWEA	Swedish Work Environment Authority
SWEM	Systematic Work Environment Management
TCO	Swedish Confederation of Professional Employees
TUC	Trades Union Congress
TURI	Toxics Use Reduction Institute
TWHSLC	Toronto Workers' Health and Safety Legal Clinic
VOSA	Vehicle and Operator Services Agency
WE	Work environment
WHO	World Health Organization
WSIB	Workplace Health and Safety Agency

Preface

In the Baywood series *Work, Health, and Environment,* the conjunction of topics is deliberate and critical. We begin at the point of production—even in the volumes that address environmental issues—because that is where things get made, workers labor, and raw materials are fashioned into products. It is also where things get stored or moved, analyzed or processed, computerized or tracked. In addition, it is where the folks who do the work are exposed to a growing litany of harmful things or are placed in harm's way. The focus on the point of production provides a framework for understanding the contradictions of the modern political economy.

Despite claims to a post-industrial society, work remains essential to all our lives. While work brings income and meaning, it also brings danger and threats to health. The point of production, where goods and services are produced, is also the source of environmental contamination and pollution. Thus, work, health, and environment are intimately linked.

Work organizations, systems of management, indeed the idea of the "market" itself, have a profound impact on the handling of hazardous materials and processes. The existence or absence of decent and safe work is a key determinant of the health of the individual and the community: what we make goes into the world, sometimes improving it, but too often threatening the environment and the lives of people across the globe.

We began this series to bring together some of the best thinking and research from academics, activists, and professionals, all of whom understand the intersection between work and health and environmental degradation, and all of whom think something should be done to improve the situation.

The works in this series stress the political and social struggles surrounding the fight for safer work and protection of the environment, and the local and global struggle for a sustainable world. The books document the horrors of cotton dust, the appalling and dangerous conditions in the oil industry, the unsafe ways in which toys and sneakers are produced, the struggles to link unions and communities to fight corporate pollution, and the dangers posed by the petrochemical industry, both here and abroad. The books speak directly about the contradictory effects of the point of production for the health of workers, community, and the environment. In all these works, the authors keep the politics front and foremost. What has emerged, as this series has grown, is a body of scholarship uniquely focused and highly integrated around themes and problems absolutely critical to our own and our children's future.

Acknowledgments

This book was developed from a seminar organised by the Cardiff Work Environment Research Centre (CWERC) and held at the School of Social Sciences, Cardiff University in January 2011 to celebrate the contribution of Theo Nichols to the sociology of health and safety at work. We are grateful for the financial support for this seminar, provided through sponsorship of CWERC by the Institution for Occupational Safety and Health (IOSH) and to all those who participated. We are especially grateful to Victoria Parkin for her help in preparing the manuscript of this book, including producing some of the figures and tables and the final text in this volume.

INTRODUCTION

Safety or Profit? Critical Aspects of Governance of the Work Environment in Contemporary Capitalism

David Walters and Theo Nichols

Every year hundreds of thousands of people die as a result of work-related experiences and exposures. Despite regulatory efforts and advances in medicine and safety engineering, the global burden of work-related death, injury, and disease remains huge; it is far in excess of that arising, for example, from armed conflicts. The ILO has estimated that around 2.3 million workers die every year as a result of work-related ill-health and injury. Of these, 350,000 deaths are attributed to accidents at work, while the rest are caused by occupational ill-health. It further estimates that there are an additional 264 million nonfatal accidents each year and 160 million people with work-related illnesses. The economic cost of this loss is thought to be equivalent to 4% of global GDP or $1.25 trillion (Hämäläinen, Takala, and Saarela, 2005).

Not surprisingly, most of this work-related carnage occurs in developing and newly industrializing countries. It is equally the case that as globalization develops, much of the risk of the hazardous work formerly conducted in the advanced economies has been transferred to developing and newly industrializing countries, where the structures and resourcing for the protection of labor are often poorly developed. This notwithstanding, the figures for work-related harm in more advanced economies are also disturbing. For example, in the UK, despite over 200 years of regulation on health and safety at work, and with one of the better records internationally, there are still over 200 fatalities from incidents occurring at work annually, and the death toll from work-related disease is thought to be at least one order of magnitude higher. In addition, more than 20,000 people leave employment as a result of work-related injury or illness, and more than 2 million suffer from ill-health they believe to be caused or made worse by their work. This leads to the estimated loss of more than 30 million working days each year and costs the equivalent of 1%–3% of GDP (see James and Walters, 2005: xi). The same scale of occurrence is found in other advanced economies, and in many, it is much higher than in the UK (HSE, 2010a).

Overall, therefore, the problem of work-related harm remains substantial and significant, yet the vast majority of these deaths, injuries, and ill-health are avoidable and preventable. Moreover, there is no great mystery surrounding their cause, which is primarily the result of failures in the discharge of the legal duties held by employers to protect the health and safety of their workers. It clearly begs the question, Why is this so?

To seek an answer, intuitively, one might first look to the reasons for the failure of employers to discharge their duties and perhaps question the efficacy of regulatory surveillance in this respect. However, when policymakers and their advisers in contemporary advanced market economies consider the problem, they rarely recommend the imposition and enforcement of stricter regulatory requirements. Nor do they suggest means to give workers greater voice to represent their interests in protecting their health, safety, and well-being, despite the well-documented "union effect" on improving occupational health and safety (OHS) outcomes. Indeed, they pointedly eschew such approaches. They argue that what is needed is to encourage employers and their individual workers to engage more proactively with self-regulatory safety management, which they further claim to be to the mutual interest of all concerned.

These policy approaches are, of course, entirely in keeping with the wider political aims of the current neo-liberal project to provide the most optimal conditions in which capital may flourish globally. This requires the removal of as much resistance as possible to its freedom, whether it is caused by the "regulatory burden" it is claimed to suffer or by the attempts of organized labor to protect the interests of workers. It is equally the case that the parallel "withdrawal of the state" through its reduced resourcing serves to weaken its regulatory and enforcement capacity. Meanwhile, evidence of the success of the now increasingly favored, largely voluntary, self-regulatory market-based methods remains scant indeed.

The appropriateness of approaches to the governance and regulation of the work environment and the protection of workers from the risks of their work in advanced market economies remains a contested territory. We think so-called new policy approaches to the governance of the work environment and the theorizing on which they are based, offer little in the way of real protection for the health, safety, and well-being of workers in contemporary forms of work. It is this territory that we explore in the following pages. We will do so by trying to bring a social science perspective to our analysis, rooted in an informed theoretical understanding of the social relations of production in increasingly neo-liberal political economies. First, we remind the reader of a very old debate concerning the relation between safety and profit.

SAFETY OR PROFIT?

This book is called *Safety or Profit* because it develops some particular arguments about the relationship between the way work is governed and organized in contemporary societies and the consequences this has for the safety, health, and well-being of the workers involved. Its aim is to question the assumptions behind the identity of interest shared between labor and capital that is claimed by policymakers in this field. It is also deliberately so named, by way of acknowledging a pamphlet written by one of us in collaboration with Peter Armstrong nearly 40 years ago (Nichols and Armstrong, 1973), which contested the same set of assumptions concerning what was beneficial to protecting the health and safety of workers, which had been the *leitmotif* of the Report of the Inquiry into Health and Safety at Work (Robens, 1972). This report, commissioned by the British government, chaired by Lord Alfred Robens and set to become the most influential of such reports on regulatory reform on health and safety at work in the UK and elsewhere during the 20th century, essentially argued not only that there was a shared interest between employers and their workers concerning health and safety, but that the prevalence of injury and ill-health was the result of "apathy" and that the solution was the encouragement of greater voluntary self-regulation. In the eponymous pamphlet to which our present title returns, drawing on evidence from conversations with workers in the UK and set against a sociologically informed analysis of what actually occurs during the labor process, its authors challenged the assumptions on which these British and subsequently other national and international regulatory reforms on health and safety at work were to be largely predicated.

Since that time, as we outline in this introduction, those assumptions have been increasingly embedded in approaches to the governance of work and the protection offered in the process to the health, safety, and well-being of workers. They remain largely unquestioned by policymakers and equally untested in terms of rigorous analysis of their effects.

At the time of the Robens Report (1972) and the reforms to which it led in the shape of the 1974 Health and Safety at Work Act, when we and several of the other authors who appear in these pages first began to engage in regulatory debates, the UK and other industrialized economies in Europe and elsewhere were approaching the zenith of the period of postwar compromise. The strength of organized labor was at its peak, and the consequences of this for the form of social democracy, the nature of the state, and the extent to which organized labor was engaged and embedded in its institutions and operation was much in evidence. It was therefore a time in which reform of regulatory means of social protection might be anticipated. Indeed, this occurred on quite a major scale in relation to the protection of worker health and safety, not only in the UK but internationally, the 1970s now being widely regarded as the watershed from which current approaches to regulating OHS emerged (Walters, Johnstone, Frick, Quinlan, Gringras, and Thebaud-Mony, 2011: 28–23). At the same time and largely for the same reasons, it was a period in which there was opportunity for politicization of the discourse on work and health,

4 / SAFETY OR PROFIT?

a raising of consciousness about many of the issues at stake, not only among academics but also among political and other activists and when fundamental thinking concerning the wider meanings of the possibility of reform was much in evidence. Perhaps most significantly, this period saw a connection being forged by critics between the traditionally insular system surrounding the regulation of health and safety at work and a wider discourse concerning social inequalities in health, the acceptability of risk, and the nature of evidence as well as with the role of power in capitalist political economies.

It is with these issues that the present volume is concerned. Times have changed since the 1970s, and with them, features of the political economy, the structure and organization of work and labor markets, the balance of power between labor and capital, and the nature of governance in what is now a globalized economy. Nevertheless, the balance between safety and profit in the employment relationship in capitalist economies remains as relevant now as in the past. As the authors of the present volume argue, rigorous questioning of the assumptions behind present-day discourse on regulating health and safety at work also continues to be as important in understanding the exploitive nature of capitalism today as it was 40 years ago; arguably even more so, because the changes that have occurred since that time are the very ones that have eroded much of the gains made by labor that were in evidence some 40 years ago.

The following chapters develop these ideas more fully. We begin this introduction to them with some further reflections on the impact of the Robens Report (1972), and the assumptions behind it, before outlining the structure and content of the rest of the book.

THE ROBENS REPORT AND THE CONTINUITY OF ITS ASSUMPTIONS

It has been 40 years since the publication of the major 20th century report into health and safety at work in the UK. The Robens Report of 1972, "Safety and Health at Work," was assessed by Mr. Dudley Smith, the Conservative Under-Secretary of State for Employment, as "one of the great social documents of our time" and it was claimed that it "has provided occasion and impetus for a major advance" (Hansard, 1973). However, in its fundamentals, the Report in fact departed little from the traditional British approach to health and safety regulation. For instance, 40 years before Robens, the Chief Inspector of Factories and Workshops had opined, "The main functions of the Inspector today is instruction (on matters within the law) and advice (on matters outside the law), rather than regulation" (Home Office, 1933: 9). Considerably before this, in the 19th century, the very first Chief Inspector, Alexander Redgrave, had declared that

> we should simply be the advisers of all classes, that we should explain the law, and that we should do everything we possibly could to induce them to observe the law, and that a prosecution should be the very last thing that we should take (cited in Rhodes, 1981: 63).

The reaffirmation of these long-standing themes in the 1972 report constituted anything but a "major advance." And the report did not hold back on such matters: excessive resort to legal sanctions was considered "counter to our philosophy"; the provision of impartial advice and assistance was to constitute the "leading edge" of the Inspectorate's activities; the criminal law was seen as "largely irrelevant" (Robens, 1972: paras. 255, 161).

In retrospect, it must be admitted that in one particular respect, the Report was indeed ahead of its time. For 40 years later, the nostrum of self-regulation has gained considerably in popularity in ruling circles as part of the neo-liberal resurgence. Even here, though (and as familiar as they are to us today), Robens' assertions— "We need a more effectively self-regulating system" (1972: para. 41) and that "Our present system encourages rather too much reliance on state regulation, and rather too little on personal responsibility and voluntary, self-generating effort" (para. 28)—can themselves be traced back to a much earlier age of laissez faire. What was different about the Report, in the context of the postwar period, was the emphasis that it placed on two related ways forward: first, the reduction of statutory regulations and second, the advocacy of general duties of care, supported by voluntary codes of practice. This shift from a mandatory prescriptive approach to a process-based approach was not to be founded on the basis of internal regulation by workers. According to the Robens philosophy, "There is a greater natural identity of interest between 'the two sides' in relation to safety and health problems than in most other matters" so there was "no legitimate scope for 'bargaining' on safety and health issues" and "statutory provision requiring the appointment of safety representatives and safety committees might be rather too rigid, and more importantly, rather too narrow in concept" (paras. 66, 69).

In the UK, the Robens Report (1972) led to a certain rationalization of organizational and legal provision. It led to the establishment of general duties of care supplemented by voluntary codes of practice; and to an extension of the legal framework for health and safety at work to a further 7 or 8 million people. It also helped a move toward the unification of hitherto separate Inspectorates, and it led to the creation of a tripartite Health and Safety Commission, which brought employers and trade unions as well as the state together (an arrangement that, remarkably, outlived the Thatcher years but operated over and above workplace level). There was, however, more tidying up and continuity in the Robens Report than there was radical change, and the same applied to the 1974 Health and Safety at Work Act that followed it. Radical change in the regulation of health and safety came only later with the 1977 Safety Representative and Safety Committee (SRSC) Regulations, which, unlike the previous regulatory reforms, were expressly *not* recommended by Robens, but were testimony to the ability of British labor to advance its interests in the early 1970s. Such changes conferred on union-appointed safety representatives various rights of representation, investigation, and inspection of the workplace; rights to require that safety committees be set up; and the right to time off with pay to carry out their functions and to undergo training approved by unions or the Trade Unions Congress (TUC). Unfortunately, it has subsequently become clear that these rights have never been fully implemented in many British workplaces (Walters and Nichols, 2007: 41–114); that they simply do not apply to many workplaces, which are not unionized; that they have therefore become considerably less influential than was hoped for them as the extent of unionization of British workplaces has progressively declined since their implementation (Walters and Nichols, 2009: 19–30) and the militancy of unions within the workplace has declined.

The decline in trade unionism has been a major change in the 4 decades since the Robens Report (1972) both in the UK and in other once highly industrialized countries. The reversal in the extent and power of organized labor has been partly a function of profound economic restructuring and partly a consequence of the anti–trade unionism and legal reform that is a fundamental part of the neo-liberal project. It makes it pertinent to enquire into how those who work in non–trade union workplaces can best be protected.

Labor-market stratification has always existed on the basis of both skill and gender. Most glaringly, men and women are still found in different jobs with different exposure patterns. The expansion of the service sector (and with it, an increase in women at work), as well as the widespread intensification of labor in all sectors, has brought with it increased exposure of both men and women to previously unstudied physical and psychosocial health hazards, including ones arising from organizational constraints.

Social scientists have increasingly been concerned with the vulnerability of those in small workplaces and of those in contractually defined precarious situations—contracted-out work, home work, self-employment, and agency labor. Indeed, it is partly in response to these developments that the last 40 years have seen a growth in critical social scientific research into health and safety in industry generally. This is in some contrast to the situation at the time of the Robens Report (1972), when the field had been largely the preserve of the doctor, lawyer, engineer, ergonomist, occupational hygienist, physiological and sometimes other psychologists (Nichols, 1997: chs. 2, 4, 5). In another development however, health and safety at work is no longer confined to the provision of mechanical devices, guards for machinery, protective equipment for workers, and so on but extends to the provision of advice to employers and health and safety agencies; something which has now become an industry in its own right, populated by consultants who both individually and as the servants of large corporations are generally much given to speak of "culture."

This talk about a safety culture is often as superficial as Robens' (1972) own penchant for invoking the importance of "attitudes" or of what an earlier generation referred to as the "human factor" (a term that can still be heard today despite long-established criticisms; Nichols, 1997: 117), and it is often based on a similar lack of evidence for the remedies offered. By contrast, critical social science approaches eschew the reduction of culture to attitudes and focus on culture in the (proper) sense of how things are done. It examines how things are done under determinate conditions. It also takes evidence seriously, which is in some contrast to the Robens Report, much of which rested not on evidence as ordinarily understood but (as noted by in the pamphlet *Safety or Profit* [Nichols and Armstrong, 1973: 5]) on the statements of apparently authoritative individuals and institutions, not least employers' organizations. As Nichols and Armstrong noted however, at that time, usable evidence was very thin on the ground, and most sociologists simply ignored occupational injuries and job-induced illness, so that in reading industrial sociology books, it was most exceptional to come across any accounts of such occurrences.

The growth of interest among social scientists in health and safety is particularly to be welcomed at the present time, when an important ideological component of neo-liberalism is precisely an attack on what, in the UK at least, is unfailingly referred to as "health and safety madness". Many of the much-cited examples of this turn out, on investigation, to be either without any foundation or to apply to very exceptional cases. Sometimes they turn out to be the result of defensive actions by public officials who fear falling victim to civil litigation claims by the very entrepreneurial moneymaking agents that neo-liberals so celebrate as the key to the nation's economic prosperity. After all, the so-called compensation culture is commercially driven. But then, to an important extent, neo-liberalism is a free-floating ideology. So much is this so, that although governments repeatedly invoke "red tape" as a "burden on business" (the New Labour government in the UK having even brought into being a department of Business Enterprise and Regulatory Reform), businesses themselves sometimes seem less troubled than this emphasis would imply. A recent international study by LSE's Centre for Economic Performance, using indicators developed by McKinsey & Company, concluded that in the United States, France, Germany, and the UK, employment laws and regulations as well as trade unions were rated by managers as considerably less likely to constitute major obstacles to improving management than being able to hire managers and others with the right skills. Equally surprising from the standpoint of 21st century conventional wisdom, employment laws and regulations as well as trade unions were judged less of a major impediment in the UK than in France and Germany (Homkes, 2011). A larger point should not be lost to sight however, and it brings us back to the point of departure for the present volume. This is that charges to the effect that "health and safety madness" or the "burden of red tape" impedes efficiency and profit distract attention from the possibility that profit can impede health and safety. This is not least so, of course, when capital is strong and labor weak, an equation that invites corner cutting. Sometimes, this possibility that safety will not unerringly be put first is implicit in the arguments advanced by those who caution against increased regulation. For instance, in opposing the development of specific directors' duties in health and safety regulation, the Confederation of British Industry (CBI) protested that "specifying duties in one area of corporate activity [health and safety] may prejudice the flexibility of directors to act properly and proportionately in other [profit related?] areas" (CBI, 2009: para. 10).

The answer to the question of what is proportionate is itself compromised by the prior and widely broadcast assumption that health and safety culture is "over the top" and part of "a stultifying blanket of bureaucracy, suspicion and fear";

8 / SAFETY OR PROFIT?

such that "trainee hairdressers are not allowed scissors" and "children are made to use goggles . . . to play conkers"—all comments made by a soon-to-be British Prime Minister David Cameron in 2009. Cameron continued to trot out such comments once he became Prime Minister, declaring earnestly to the faithful at the 2011 Conservative Party Conference, "One of the biggest things holding people back is the shadow of health and safety" (Cameron, 2011) and topping these comments in his New Year speech in 2012, in which he announced his government's "clear New Year's resolution to kill off the health and safety culture for good" (Cameron, 2012: 6).

THE STRUCTURE AND CONTENT OF THIS BOOK

Several features alluded to in the previous section concerning the assumptions behind the Robens Report (1972) are explored in greater depth in the following chapters.

In Part I of the book, we consider the current structure and organization of work and the labor market in advanced market economies and their implications for the health, safety, and well-being of workers involved. Enormous changes have taken place in recent decades, and in Chapter 1, Michael Quinlan presents an overview of their impact on OHS with special reference to research on the growth of precarious employment since the mid-1980s. He argues for a greater acceptance of the importance of work organization, broadly defined to include relations of power, as against loose notions about organizational culture, as essential to a proper understanding of the consequences for both the health of workers affected by these changes and for the means to ameliorate such effects.

One of the most significant features of change accompanying the restructuring of work in former highly industrialized economies has been the changing gender balance of the labor force, with increasing numbers of women now employed not only in the service sector but also in sectors in which they were previously absent. This has been accompanied by increasing numbers of mothers continuing to be active in the labor market. In Chapter 2, Katherine Lippel and Karen Messing present a gender perspective on the governance and regulation of the work environment in which they review changes in both the regulatory frameworks governing women's work and in the role of women in the labor force itself. They examine the hazards specific to the kind of work in which women are disproportionately represented, as well as the comparative precariousness of their employment arrangements and its impact on their health, safety, and well-being. They outline the nature of the challenge this presents for governance and regulation of the organization of work and the work environment, and consider how successfully it has been met. They find strong evidence of regulatory failure in this respect and discuss some of the main reasons for this failure.

In Part II, attention shifts to a more detailed examination of the governance and regulation of the work environment under the influence of neo-liberal economic and political strategies, as evidenced in a variety of national and international settings.

In the previous section, we argued that in the UK, the most significant departure from British tradition in terms of the governance and regulation of the work environment during the 1970s was not the recommendations of the Robens Report (1972) and the measures to achieve its recommendations, introduced under the HSW Act, but rather the result of regulations later issued under the same Act, conferring rights on trade unions to appoint workplace health and safety representatives. These provisions were rooted in campaigns by the trade unions and not in recommendations of the Robens Committee. We have written elsewhere on the fate of these arrangements under successive British neo-liberal governments (see, for example, Walters and Nichols, 2007; Nichols and Walters, 2009).

In Part II, we begin by examining the resilience (or otherwise) of such arrangements in Sweden, where measures on managing the work environment have been widely regarded as significant examples of progressive provisions and practices that especially emphasize a participatory approach. Perhaps more than anywhere else, these measures have represented significant and substantial efforts to facilitate the involvement of workers and trades unions with work environment issues at the workplace and beyond. Nor were such measures restricted to larger organizations in Sweden, as is the case in practice in many other countries, but included specific focus on achieving trade union representation on health and safety matters for workers in smaller firms too. The strength of the labor movement in Sweden and the extent of the penetration of trade union membership even to the smallest workplaces was an acknowledged reason for these progressive approaches. Yet, in recent decades, Sweden has experienced much the same reorganization and restructuring of work as in other European countries, while more recently, it has also been subject to the effects of a strongly neo-liberal government. The consequences of these changes for the so-called Swedish model are explored in Chapter 3, wherein Kaj Frick looks especially at the situation of Swedish trade union health and safety representatives and the extent of their continued ability to influence OHS in the context of the demise of social democratic government, the weakening of labor, and changes in the labor market.

The second chapter in Part II turns attention to the governance of OHS by means of a critical examination of the nature of so-called new governance theory. It observes the extent to which "new governance" has become a booming academic industry offering a theoretical understanding of governance under neo-liberalism and claiming something more for it than mere deregulation or neo-liberal self-regulation. As such, it has found favor in many spheres of current regulatory policy.

Reviewing the emergent and mainly U.S.-based literature on new governance in relation to OHS, Eric Tucker makes the important point that in the area of occupational health and safety, new governance is hardly new at all. Indeed, he argues that in this field, what are now labeled new governance concepts were articulated and applied in the Robens Report in 1972 and were especially evident in its critique of command and control legislation and its emphasis on the need to develop better self-regulation. In this context, he therefore finds it useful to revisit the critique originally aired by Nichols and Armstrong in *Safety or Profit* (1973), to seek some "old" lessons for new governance in OHS regulation. He notes that, as argued 40 years ago, a proper understanding of risk creation needs to take as its starting point the pressure for production that is generated within capitalist relations of production. From a political economy perspective therefore, the central regulatory problem was and remains how to counteract this pressure to prioritize production over safety. Using the example of the Canadian province of Ontario, Tucker discusses the roots of new governance in the Robens Report and briefly reviews Ontario's experience of its dynamics and its vulnerability to regress toward neo-liberal self-regulation/paternalism in the absence of effective worker OHS activism. He finds little comfort to be found in the means to prevent this through prescriptions embraced by the most prominent theoretical proponents of new governance and argues that these are unlikely to be institutionalized with the protective conditions they advocate and that their emphasis on self-regulation encourages a movement toward the destination they claim they wish to avoid. Lastly, he asks whether such a degradation toward neo-liberal self-regulation/paternalism is inevitable and if not, whether a progressive new governance theory is possible and has anything to offer toward strengthening a regime of public regulation under the unfavorable conditions that prevail today.

Chapter 5 returns us to the UK, where, as Steve Tombs and David Whyte describe, governance of health and safety at work has been subject to significant redefinition and redirection in recent times. They take as a starting point the safety/profits couplet: Nichols' and Armstrong's (1973) pamphlet demonstrated how safety and profitability ultimately stand in contradiction, so that better safety may only be won at the expense of levels of profitability and that this in turn is only likely to be achieved, in the main, through greater power for workers over the conditions of their exploitation. In their chapter, Tombs and Whyte examine how, in the era of neo-liberalism, the common sense of safety and profits has been refashioned, under-pinning a new emphasis in the approach to regulatory policy and enforcement. Attempting to explain how such trends have been rolled out, embraced not least by the Health and Safety Executive (HSE) itself, they indicate how the business case for safety—a claim for safety *and* profits—along with a whole series of more or less familiar related "concepts," have been dusted off and given considerable refurbishment. They argue that this new version of common sense about safety and profits (in which safety appears as an impediment to profit, which is depicted as the greater good) is something that the British trade union leadership has failed to confront effectively. In so doing, they pay particular attention to the last decade and the record of the then-Labour Government toward the regulation of safety and health at work. Setting out recent trends in safety and health law enforcement, they demonstrate that every area of the HSE's formal enforce-ment activity declined significantly during the first decade of this century. Finally, the chapter raises fundamental questions about the role of the trade union movement in defending worker safety as we enter a period in which those attacks on basic forms of regulatory protection are set to intensify. This chapter stands as a striking vindication of Nichols and Armstrong's analysis in the contemporary politics of regulation, while also emphasizing the distinctiveness of the contemporary common sense.

In a similar vein in Chapter 6, Richard Johnstone examines the approach to health and safety regulation in different Australian jurisdictions. He begins by reviewing the history of regulatory approaches to OHS, where, until the 1980s, the various Australian OHS regulatory agencies uniformly adopted the "advise and persuade" approach to enforcement, with prosecution as a last resort. From the early 1980s, however, there were, at least in some of the Australian jurisdictions, attempts to strengthen state enforcement by increasing maximum penalties, exploring various models of imposing liability upon corporate officers, empowering inspectorates to issue on-the-spot fines, and recently, enabling inspectorates to accept enforceable undertakings and providing courts with nonpecuniary penalty sentencing options. Other developments involved empowering employee representatives to play an enforcement role by enabling union-initiated prosecutions and empowering health and safety representatives to issue provisional improvement notices and to direct that dangerous work cease. Johnstone examines the evidence of the use of these approaches and locates them within a wider discussion of the recent decriminalization of OHS by drawing on a range of academic and policy debates on enforcement to map out the complexities of, and assess, OHS enforcement in Australia.

In Part III, we turn attention to the examination of assumptions implicit in the processes of governance of OHS. Our concern here is fundamentally with the role and limitations of the use of evidence in decision making on OHS strategies at state and company levels. Once again, our analysis takes as its starting point conclusions put forward by Nichols and Armstrong (1973) concerning what passed for "evidence" in the Robens Report (1972), which they saw to frequently consist of statements from apparently authoritative individuals and institutions rather than constituting what is ordinarily understood by the term, let alone what critical social science may require of it. We do this by examining some examples of what has informed decisions made during the processes of governance of contemporary occupational and environmental health issues and through case studies of the operation of regulatory systems in situations of trade liberalization and organizational restructuring.

Occupational diseases remain a relatively neglected low priority for both research and policy despite their public health importance. Chapter 7 examines why this is so and what is being done to address the neglect. It includes a case study of occupational cancer and shift work, including night work, using UK and European research and policy responses, to illustrate the interplay between competing interests and to highlight the different responses within and beyond the UK to the challenge of recognition and prevention of occupational cancer.

Andrew Watterson analyzes the toll taken by occupational cancer in the UK and globally. He examines how different types of evidence are weighted and contested in the policy process and the current and historical role of employers, scientists, politicians, regulators, trade unions, and workers in addressing or downplaying the

12 / SAFETY OR PROFIT?

problem. Cancer, along with other occupational diseases, remains a relatively neglected research and policy priority despite its public health importance. A case study is used of occupational cancer, shift and night work, drawing on UK and European research and policy responses, to illustrate the competing interests and highlight the different responses within and beyond the UK to occupational cancer recognition and removal. Historically, bodies such as Science for People and particular trade unions identified and campaigned for action on UK occupational cancers linked to work with trade unions. Worker testimony on and concerns about occupational cancers are often ignored, and the reasons for this are documented. The international complexity of the position is reflected in how some governments and regulatory bodies act on or oppose international research findings on specific occupational cancers or reflect employer apathy, sometimes trade union apathy, and scientific complacency about the subject. This may be compounded by "soft" regulation and/or no enforcement on many occupational cancers. Such responses ensure that capital rather than civil society concerns win out. The chapter shows, however, that by using research from WHO bodies and academic institutions, trade unions and civil society bodies have begun to work out preventative strategies to address the social, human, and economic costs of occupational cancers, both at an international and UK level.

In Chapter 8, Wayne Lewchuk returns us to the Canadian province of Ontario with a case study that sets out to explore labor, management, and government concerns regarding the structures and procedures of the system for the governance and regulation of health and safety that have emerged over the last 3 decades and assesses views on how they must adapt in the face of restructured labor markets and the spread of precarious employment. He goes on to present a case study of a recent Canadian review of health and safety regulation. In January 2010, the Province of Ontario announced the formation of a panel to review the effectiveness of health and safety regulation in the province. The numerous submissions by employer groups, unions, and other interested organizations provide a rare window into different views on how well the system is functioning and the changes that are needed to adapt to the spread of precarious employment. The Review Panel's final report, released in December of 2010, shares insights into the future direction of health and safety regulation in Ontario and how one province proposes to deal with the concerns raised by the spread of precarious employment. The focus of Lewchuk's analysis, however, is the process by which these insights and recommendations were arrived at. Using an analysis that is situated within and informed by essentially the same theoretical framework as that of Nichols and Armstrong (1973) in their critique of the assumptions behind the Robens Report (1972), he considers how the evidence presented to this more recent inquiry was treated.

In Chapter 9, Charles Woolfson considers the causes and circumstances of the loss of the *Deepwater Horizon* oil platform, the explosion and oil spill that occurred in the Gulf of Mexico in 2010. He shows that a lethal cocktail of contingent circumstances and systemic underlying causes contributed to the disaster. These included multiple safety systems that did not function at crucial moments, managerial failures both in the immediate run-up to and during the unfolding disaster, organizational failures that were embedded in distorted information flows and the lack of coherence of safety management systems, defective regulatory oversight by authorities with contradictory responsibilities for both production and safety, and the outright "capture" of regulatory processes by the industry itself. Woolfson suggests that a simple focus on the firms involved and their widespread portrayal as "corporate villains" obscures a deeper question. How is it that the safety regime in the U.S. offshore industry, in the very heartland of the global oil industry, seemingly remained impervious to the lessons of previous disasters? That this disaster was avoidable is true also in retrospect of every other disaster. That it was foreseeable is equally so. The more fundamental question for Woolfson is therefore *why* has safety remained an intractable issue in the offshore industry? In providing their answer, he points to a previous study of offshore safety by W. G. Carson (1982), which spoke tellingly of a "political economy of speed" and "institutionally tolerated non-compliance" in which profits are prioritized over safety, not only by the companies concerned but also by the desire of governments for lucrative oil revenues. Woolfson argues that today, the legacy of the long era of deregulation in the United States, exported globally in the name of free enterprise and economic efficiency, continues to take its toll in worker lives and the well-being of communities. He suggests that it remains to be seen whether this unpalatable truth will find acceptance by those with the power to reconfigure safety regimes to protect people first and profits second. And he concludes that appalling though it may be to contemplate, unless this recognition is translated into action by public policymakers, other *Deepwater Horizons* will happen.

Most of the chapters that are outlined above and detailed in the rest of this book were first aired as papers at a conference held in Cardiff in 2011 to celebrate Theo Nichols' contribution to sociological thinking about work and health in capitalist economic systems. As that rather grey winter day in the Welsh capital unfolded, continuity between the thinking behind the various speakers' presentations and coherence in their analyses became increasingly apparent. Reviewing the day, we felt that their collective efforts amounted to a significant contribution to the sociological understanding of issues behind the regulation of the work environment over the past 50 years or so and worth elaborating in their present form. The speakers on that day, like the chapters in this book, concerned themselves with current issues, but the more universal salience of theoretically informed inquiry they each brought to their subjects, in the end, amounted to a framework for understanding some of the weaknesses in the assumptions behind current governance of health and safety at work. More than this though, it offered a way of conceptualizing what lies beneath current discourse on the governance and regulation of the work environment of millions of people and a robust challenge to the pervasive "conventional wisdom" that dominates current neo-liberal thinking.

We think the final versions of the papers first discussed in Cardiff that are presented as the following chapters have succeeded in capturing something of the excitement felt by the participants on that day. For us, its legacy has been the hard work and shared sense of purpose that has driven the production of this book. We therefore hope it will provide some cause for reflection for all who participate in decisions on the governance and regulation of the work environment internationally and ultimately be of some small benefit to the workers who should be the rightful beneficiaries of such decisions. We offer some final reflections in an Afterword.

PART I

Economic Restructuring, Labor Market Stratification, and Their Consequences for Health and Safety

What happens with respect to health and safety at work is affected not only by internal relations and external regulation but over and beyond this by the wider political economy, including, among other things, growth rates, unemployment, the structure of the labor market, trade union rights, and coverage. Historically, variable types of work contracts are a constituent part of this wider context, and Part I begins with a review of the resurgence of precarious forms of work by Michael Quinlan in Chapter 1. Drawing upon research not only from his native Australia but also extensively from a range of other countries on other continents and from different historical periods, Quinlan documents a variety of way in which workers are and have been rendered insecure and the consequences of these precarious forms of work for their health and safety (and indirectly for the well-being of yet other workers). Chapter 2 then provides a gender perspective. Here, Katherine Lippel and Karen Messing present a wide-ranging discussion of the regulation of women's work over the last century, which casts light on women's occupational health, a subject that has often been ignored, including sometimes by those interested in women's public health who are apt to forget women in their role as workers. In doing this, they too draw not only on their native Quebec but on research conducted in several different continents and countries.

http://dx.doi.org/10.2190/SOPC1

CHAPTER 1

Precarity and Workplace Well-Being: A General Review

Michael Quinlan

INTRODUCTION: NEO-LIBERALISM, FRACTURING OF THE LABOR MARKET, AND INCREASED INEQUALITY AT WORK

Over the past 4 decades, changes to business practices along with the dominance of neo-liberal policy discourse have reshaped the organization of work and the social protection framework in ways that a growing body of evidence has shown to be antithetical to the health, safety, and well-being of workers, their families, and the broader community. Stripped of euphemistic spin, the shift to "flexibility" that typifies the "new" world of work is one marked by precariousness, disorganization, and regulatory retreat as labor is increasingly treated as an expendable commodity to be traded globally like any other factor of production. The growth of precarious work provides a stark and arguably pivotal example of how workplace health and safety has been subordinated to the pursuit of profit and sectional interest. This chapter points out that precarious employment is not a new phenomenon nor is evidence of its health-damaging effects. Rather, it is a problem that has reemerged with the rise of neo-liberalism. This chapter points to what can be learned from historical experience as well as recent attempts to explain how precarity damages health (including the economic pressure, disorganization, and regulatory failure, or PDR model).

Since the mid-1970s, there have been a number of profound global changes to labor markets and work organization. These changes were integral to the abandonment of the postwar Keynesian policy accord that operated in most rich countries. In its place, there was the reassertion of a "market-based" policy framework that eschewed collective organization of workers, public sector economic activity, and social protection/state intervention (except to privilege private property and its owners) that came to be termed *neo-liberalism*. The shift was linked to and interacted with a renewed assault by capital (and especially some elements within it such as the finance sector) on organized labor and the "restrictions" it imposed on labor-market flexibility. The changes to work organization include repeated rounds of organizational restructuring through downsizing, outsourcing, and privatization often in association with the increased use of extended supply chains (global and nationally) that rely on elaborate networks of subcontracting. For workers, the effects of these changes has overwhelmingly entailed work intensification and less employment security as capital sought to extract ever-greater surpluses from a given input of labor (with fancy titles such business process reengineering and lean production barely disguising the central object of cost cutting, including labor costs). It has also impacted the structuring of work arrangements in terms of legal status and entitlements. Notable here (especially for old-developed rich countries) has been a growth (more accurately, a reemergence to prominence) of more precarious or contingent work arrangements, multiple job-holding, and long or irregular working hours.

As studies increasingly show, the shift to more flexible (flexible for whom?) work also impacts the work intensity and security of even those who retain nominally secure employment. The aging of the population in many rich (and some poor) countries and the increased use of foreign workers (not just migrants) has added layers of complexity and vulnerability to the growth of precarious work. As a result, the changes just described impact the vast bulk of the working population. As will also be shown, the effects of these changes are not only profound for workers but also extend to their families and the community more generally.

A series of terms has been used to describe these changes in work arrangements, most notably *contingent work* and *precarious employment*. This chapter uses the term *precarious work* rather than *precarious employment* because work is a broader term, and a substantial number of those found in these arrangements (such as subcontractors and many homecare workers) are not engaged under an employment relationship.

The aim of this chapter is to review the current state of knowledge about how precarious work impacts the health, safety, and well-being of workers and others in the community. It is divided into four parts. The first puts the current debate over precarious work and its health effects into historical context. The next two sections then look at contemporary evidence on the health effects and regulation/ social infrastructure, respectively. The fourth and final section then examines a number of attempts to explain the mechanisms by which precarious work arrangements damage health.

PUTTING PRECARIOUS WORK AND ITS EFFECTS INTO HISTORICAL CONTEXT

What is often termed the "new" world of work is not especially new. Indeed, what was termed or seen as the standard employment (relatively secure full-time work) was only really dominant among wealthy countries for a period of around 50 to 60 years in the middle quarters of the 20th century. It was the contingent result of a prolonged period of mobilization by organized labor and other progressive groups, as well as Keynesian full employment and income redistribution policies. The abandonment of the latter (and rise of neo-liberalism as the dominant policy framework for all social decisions) in the early 1970s also coincided with the beginning of the shift in business practices that promoted more contingent/precarious work arrangements such as outsourcing/subcontracting, corporatization, privatization, downsizing/restructuring, increased use of temporary workers, and various forms of "lean" management. It was not until the 1980s and 1990s that the consequences of this for patterns of employment began to be widely recognized.

However, in many respects, this marked a return to widespread precarious employment that had been pervasive in the 150 years prior to the Second World War. During this period, secure employment was the exception, with large numbers of workers holding jobs that were low paying and precarious, including sweated labor and outworkers (mainly women), child labor, casual labor (e.g., docks, agriculture, navvies), indentured immigrants (especially non-European), merchant seamen, fishermen, and an array of subcontracted labor (in construction, transport, and mining).

Indeed, the very term *precarious employment*, coined in the 1980s, was actually a reinvention, because the very same term had been used to describe temporary and insecure work in the policy sphere (such as UK House of Commons debates) on a regular basis since the early 19th century until the 1930s (its demise coinciding with the rise of "standard employment"). During this earlier period, the social ills associated with precarious employment had been painstakingly documented and recognized. The final report of the UK Royal Commission on Labour (1894: 73–87) devoted a substantial section to irregularity of employment, including the various forms it took (in agriculture, the docks, and in the sweated trades, for example) and reasons for it. Among the solutions canvassed by the Royal Commission (1894: 75, 79) for "employment that was irregular and precarious" was an extension of employment by government authorities (including local government) and movement away from putting such work out to private tender. Some 14 years later, the Royal Commission into Poor Laws (1909: 223–224, 631) in the UK identified the casual labor system (i.e., temporary work) as the single most important industry-related cause of poverty/pauperism (which in turn had cascading effects on children, education, and health), identifying casual dockworkers as an archetypal case. The report canvassed various ways of securing decasualization, but no measures were undertaken until World War II and, decasualization was not achieved until the 1960s (a similar timeline applied in Australia).

Just as precarious modes of work are not new, nor should the mounting evidence of their adverse effects on worker health, safety, and well-being (discussed later) come as a surprise, since there was copious evidence linking precarious work arrangements to injury, disease, and ill-health during the first industrial revolution. Prefiguring contemporary concerns with the adverse OHS effects of precarious employment, there is actually manifold evidence of the impact of insecure work arrangements, contingent payment systems, subcontracting, and irregular or long

20 / SAFETY OR PROFIT?

working hours on worker health in the 19th century and early 20th century (Quinlan, 2011, under review). In addition to material collected by social reformers, unions, and the like, this evidence can be found in parliamentary papers, disaster inquiries, and royal commissions into working conditions in factories, mines, and shops; the writings of academics; and in medical/public health and other learned journals.

One example of this was the practice labeled as "sweating" in clothing and other trades, where the combination of low pay and long hours and subcontracting to poor health outcomes were extensively documented between 1880 and 1920. In 1888, for example, the leading medical journal *The Lancet* commissioned its own special sanitary commission into sweating, which pointed to the pervasive nature of these arrangements, the exploitive role of middlemen, and the recurring connection between low and irregular earnings; poor quality food; cramped working conditions; crowded, drafty, poorly ventilated, and dirty accommodation; filth and poor sanitation; fatigue, chronic injuries, and poor health; and susceptibility to all too common infectious diseases (such as scarlet fever), which led to a higher mortality rate among children, both those working and those not (*The Lancet*, 1888: 37–39).

Like a number of informed observers in North America (such as Florence Kelley), Australasia (such as factory inspector Charles Levey), and elsewhere, *The Lancet* (1888) emphasized the connection between subcontracting, low pay, and poor health. It also emphasized that these forms of work organization were the result of deliberate choices and open to alteration. While conditions were worst for those concentrated at the bottom of the subcontracting chain (predominantly women and children), the competitive pressure of progressive subcontracting of work had spillover effects on wages, employment security, and health, even of male journeymen tailors engaged in factories and "better" workshops. The absence of income security also eliminated any incentive for the employer to allocate work in a planned and efficient manner. Writing about journeymen tailors, *The Lancet* observed,

> The irregularity of employment and of income must be a fruitful source of disease. For instance, while there is much enforced idleness, a tailor has often to perform "nine days' work in a week." The insufficient sleep, the strain to the eyes, the lack of proper time to take meals or out-door exercise, and the prolonged confinement in unwholesome and over-heated workshops are naturally important factors in undermining the constitution of even the most fortunate among the journeymen tailors (*The Lancet,* 14 July 1888: 740).

As the reports made clear, women tailoresses and machinists were in an even worse situation, with *The Lancet* reporting a case in which a tailoress in Manchester, unable to earn a subsistence living (or repay debts for food), attempted suicide.

The Lancet pointed out how low paid and insecure work had cascading effects on families, requiring women and children to take on the most hard and poorly paid tasks. Referring to outworkers near Dudley, *The Lancet*, 2 June 1888: 1101) observed,

> Groups of girls may be seen trudging along with bundles balanced on their heads. The bundles generally contain moleskin trousers, often weigh half a

hundredweight, and have been carried sometimes for more than miles. These are the home workers, the wives and daughters of men [predominantly miners] whose earnings are insufficient to keep their families (*The Lancet*, 2 June 1888: 1101).

The Lancet also pointed to adverse effects of these precarious forms of work on public health (including the reluctance of outworkers to report infectious diseases for fear of losing work), not only the children of children of sweated labor but also consumers (often unaware of the origins of the products) and the public more generally.

> The clothes at times are contaminated, the workers so starved and exhausted that they must soon fall victims to wasting disease when they are not actually driven to suicide. This is a matter of such immediate importance, and which every sentiment of humanity is so concerned, that petty quibbles over the details of doctrinaire political economy must not be allowed to stand in the way of those sweeping and far-reaching reforms that alone can deal with the widespread evils now fully revealed to the public' (*The Lancet*, 2 August, 1890: 246).

In a statement that applies just as much today, *The Lancet* argued that irrespective of where they are produced, when goods are produced for public consumption, the public has a right to have a say in the conditions of production (*The Lancet*, 2 August 1888: 246).

The foregoing is just a small illustration of the substantial body of historical evidence linking precarious work arrangements to adverse health outcomes. In addition to sweated labor (itself by no means confined to clothing workers but found in an array of other industries), adverse health and safety effects were also well documented in government inquiries, health journals, and the like with regard to casual work such as dock laborers, construction workers/navvies, and rural/agricultural workers; fixed contract workers like seamen, whalers, and fishermen; children working as itinerant labor (in factories, homes, and as street hawkers); and own-account subcontractors such as some mineworkers in the period 1880–1930 (Quinlan, 2013, under review). Again and again inquiries and the like pointed to the health-damaging effects of long or irregular hours, low/erratic pay, and work regimes in which labor was treated as entirely expendable. As with sweating, these reports pointed to externalities or spillover effects, including the effects of low and irregular pay on diet and accommodation, the spread of infectious diseases to the broader community (especially in the case of sweated home-workers, dockworkers, and seamen), the growth of child labor (and effects on children's education), and the "burden" imposed on the community/state as older/disabled precarious workers were discarded by employers. Unfortunately, this rich vein of prior knowledge on the health effects of work organization was all but forgotten by researchers examining the "new world of work" from the 1980s and ignored by those promoting flexibility in the labor market.

REDISCOVERING THE ADVERSE HEALTH EFFECTS
OF PRECARIOUS WORK

Since the mid-1980s, a growing body of research has linked these changes—and downsizing/job insecurity, subcontracting, temporary employment, and outsourcing/home-based work in particular—to a substantial deterioration of OHS outcomes. There are now literally hundreds of published studies undertaken in dozens of countries, using an array of different methods (longitudinal, cross-sectional, case-study, etc.), study groups (general population samples, working-age cohort studies, workplace and industry-specific studies), and indices (injury statistics, hazard exposures, physiological health measures, psychological well-being, drug use, work/family balance, etc.). A series of reviews of these studies has found substantial consistency in results as to the health-damaging effects of these work arrangements; results that have been maintained over time (Benach, Muntaner, and Santana, 2007; Quinlan and Bohle, 2008, 2009; Quinlan, Mayhew, and Bohle, 2001; Virtanen, Kivimaki, Joensuu, Virtanen, Elovainio, and Vahtera, 2005).

More recent research has continued to refine and expand understanding of the complex interconnections between insecure and contingent work/payment systems and health, including drug use by contingently paid truck drivers and workers experiencing downsizing (Kivimaki, Honkonen, Wahlbeck, Elovainio, Pentti, Klaukka, et al., 2007; Williamson, 2007). Other studies point to the health-damaging effects of intermittent work and bouts of unemployment (Melanfant et al., 2007). There are also complex interactions between precarious work and working hours/work/life balance, including the health-damaging effects of presenteeism and irregular working hours, especially when workers have no control over the variability (Aronsson, Gustafsson, and Dallner, 2000; Bohle, Willaby, Quinlan, and McNamara, 2011; Dembe, Erickson, Delbos, and Banks, 2005; Simpson, 2000). Important gender dimensions have also been identified. In a recent study of how nurses with atypical work schedules reconcile family responsibilities based on 24-hour observation, Barthe, Messing, and Abbas (2011) found that for workers with heavier family responsibilities, choice of work schedules was almost entirely conditioned by family considerations, leaving little leeway to manage workers' own health protection. Recent research also suggests perceptions of organizational justice have effects on work/life balance (Elovainio, Kivimaki, Linna, Brockner, van den Bos, Greenberg, et al., 2010).

There have also been efforts to place the health effect of changes in work into a global context. As part of a World Health Organization (WHO) initiative on the social determinants of health—or more accurately, health inequalities—the EMCONET group produced a report (Benach et al., 2007) on employment-related health inequalities that extensively documented the extent/trends and health effects of four specific arrangements, notably precarious employment, the informal sector (which accounts for over half the workforce in many poor countries), child labor, and slave and forced labor. This report (and subsequently, an extended version published in book form; see Benach et al., 2010) provided both a global perspective and one that balanced the preoccupation with precarious employment in rich countries with the pernicious effects of the large/growing informal sector as well as child and forced labor in poor countries (though these are also found to varying degrees in rich countries). The parallels between the essentially unregulated informal sector of poor countries today with very similar situations in rich countries just over a century ago should not be ignored. For example, a Brazilian study (Iriat et al., 2008) highlighted the importance of the absence of legislative protection in shaping the attitudes of informal workers toward occupational health hazards in their jobs.

In a further disturbing echo of the past, a growing body of research in both rich and poor countries points to the cascading effects on community and public health of work that is poorly paid, intermittent/insecure, and associated with long or irregular work schedules. These include impacts on households due to difficulty budgeting, housing quality, effects on children's education and health, diet/obesity (Aronsson, Dallner, Lindh, and Goransson, 2005; Barling and Mendelson, 1999; Buxton, Quintiliani, Yang, Ebbeling, Stoddard, Pereira, et al., 2009; Devine, Farrell, Blake, Jastran, Wethington, and Bisogni, 2009; Labonte and Schrecker, 2007). There is also substantial international evidence that cuts to hospital staffing levels have led to an increase in infections and errors (see, for example, Aiken et al., 2002; Stegenga, Bell, and Matlow, 2002). The use of subcontractors (including multitiered subcontracting) has been linked to public safety issues in road and air transport as well as disastrous workplace incidents in which health and environmental risks have extended to neighboring communities such as the explosion at the AZF fertilizer factory in Toulouse, France, in 2001 and the explosion on the *Deepwater Horizon* oil rig in the Gulf of Mexico in 2010 (Johnstone, Quinlan, and Mayhew, 2001; Loos and Le Deaut, 2002; Quinlan and Bohle, 2008; Woolfson, Foster, and Beck, 1996). Elaborate supply chains using contingent workers in food harvesting and processing may also contribute to food safety issues, but this is yet to be systematically explored.

While knowledge is expanding, the rapid growth of research into the health effects of precarious work has not systematically addressed all the key areas of labor-market change. For example, important areas of change to work organization such as outsourcing/subcontracting, privatization (for an exception, see Ferrie, Martinkainen, Shipley, Marmot, Stansfeld, and Smith, 2001), self-employment (including microbusinesses, home-based work, and franchising arrangements) have received little attention, although in some, like homecare, the number of studies is gradually expanding. A Danish study of privatization/outsourcing of bus drivers (Netterstrom and Hansen, 2000) found that outsourced drivers experienced deterioration in their psychosocial work environment, as well as a variety of physiological symptoms of stress including increased systolic blood pressure. Unfortunately, research into subcontracting and privatization appears to have largely stalled (for an exception, see Nenonen, 2011). Another major gap is with regard to supply chains—complex series of contractual arrangements designed to secure a good or service and increasingly global in character. There are only a handful of studies, although when combined with indirect evidence, these suggest significant adverse health effects (Lloyd and James, 2008; Walters and James, 2009).

Another problem is that until recently, many studies of the OHS experiences of immigrants/foreign workers, young workers, or women failed to give adequate attention to the precarious nature of the jobs in which they were typically concentrated (for a more nuanced approach about layers of vulnerability, see Sargeant and Tucker, 2009). Similarly, there has been a common failure of those studying OHS in small business to take account of the concentration of precarious workers in those organizations and that many are "bottom-tier" subcontractors. There are noteworthy exceptions to these problems just identified that appear to be gaining traction, albeit slowly (see, for example, Bohle, Pitts, and Quinlan, 2010; Lipscomb et al., 2008; Menendez et al., 2007; Premji, Messing, and Lippel, 2008; Seixas, Blecker, Camp, and Neitzel, 2008; Walters, 2001). For example, in a study of hotel workers, Siefert and Messing (2006) undertook a detailed ergonomic examination of the work of hotel cleaners, identifying the intensification of physical effort (including the number of rooms to be cleaned by each cleaner) associated with a shift to contingent work arrangements. At the same time, the study examined the gendered nature of task allocation, the failure to provide even basic protective equipment, and the additional vulnerability associated with the predominantly immigrant workforce (including some undocumented workers). There have also been efforts to propose a multitiered layering of vulnerability to both incorporate and explain the complex interaction of precarity with gender and ethnicity (Sargeant et al., 2009).

There are a number of methodological and conceptual issues that also require attention, three of which are worthy of mention. First, there has often been a failure to sufficiently differentiate within categories of precarious work (such as the diverse array of temporary workers that includes seasonal, on-call, agency, and fixed-contract workers) and between categories of precarious work (most studies compare only one or two categories of precarious work with permanent workers when studies have empirically validated over six distinct categories; Louie, Ostry, Keegal, Quinlan, Shoveller, and LaMontagne, 2006). The result is that the focus of some studies is too narrow and that care is required in both international studies and metareviews of the literature to ensure that what is compared is like with like. For example, research suggests there are significant differences between fixed-term contract, seasonal, and casual temporary workers and between direct- and indirect-hire temporary workers that can affect OHS outcomes (Schweder, 2009; Underhill, 2008). Even when categories have been differentiated in comparing them with nonprecarious workers, there is a need to take account of the length (and exposure effects) or irregularity of working hours. Second, while many studies compare precarious and nonprecarious workers in order to gauge health effects, there is evidence that the presence of precarious workers in a workplace or industry can alter the working conditions and health outcomes of nonprecarious workers over time due to the shifting of supervisory/training responsibilities onto the latter, or direct competition for work between the two groups (Mayhew and Quinlan, 2006). These might be termed spillover effects—something discussed in more detail in a later section.

Third, there is also debate as to the extent to which precarity is purely determined by employment status and whether additional or substitute measures should be developed (Lewchuk, Clarke, and de Wolff, 2008; Vives, Vanroellen, Amable, Ferrer, Moncada, Llornes, et al., 2011). For example, Muntaner, Benach and Vives (see, for example, Vives et al., 2011) have developed a multidimensional employment precariousness scale. They argue that this is superior to an approach simply based on the employment contract type, because it overcomes the problem of the heterogeneity of contract types (referred to above), accommodates changes to regulatory regimes, and is more consistent with the notion that precariousness represents a continuum with a progressive shift in the labor market occurring in this direction, which affects even those holding nominally permanent employment. Using this approach, a recent study in Spain found almost half of all waged workers (6.4 million of 13.3 million) were exposed to precariousness, and precariousness seemed to have significant consequences in terms of mental health (Vives et al., 2011: 641–642). Approaches along the lines just described enable researchers to reconcile how the impact of downsizing/restructuring (and consequent job insecurity) on permanent employees is just as integral to the study of precariousness as research on temporary workers and subcontractors. This issue is examined in more detail in the final section of this chapter, which explores the more general question of how precarity affects health.

The complexities and gaps in knowledge just identified should not detract from the broad findings and remarkable consistency of existing evidence with regard to the effects of precarity on worker health. While these gaps/complexities should be addressed, there is already enough evidence to indicate that a radical readjustment in policies relating to work and its health consequences is required. However, as the next section will demonstrate, pressure for reform has largely failed to gain traction in the context of an overarching obsequiousness to neo-liberalism among policymakers.

REGULATORY AND INFRASTRUCTURAL EFFECTS OF PRECARIOUS WORK

In the light of the evidence just described, an obvious question to arise is the extent to which existing social protection regimes are meeting these challenges or are being modified to do so. With regard to infrastructure, mention should be made of the cascading community health effects referred to in the last section, many of which will become burdens on public health and social support apparatuses. There is evidence that the increase of insecure and intermittent work brings with it increased claims on social security, including unemployment payments; poverty-based payments (for food, healthcare, children's education, and accommodation); and disability pensions, while also weakening the coverage/support afforded by systems that employers do contribute to; notably both income support and return-to-work schemes under workers' compensation (Cranford and Vosko, 2006; Quinlan and Mayhew, 1999; Underhill, 2008; Vahtera, Kivimaki, Forma, Wikstrom, Halmeenaki, Linna, et al., 2005). In short, the growth of precarious work is associated with externalities that impose a great burden on the welfare state and does so at a time when neo-liberal policies seek to wind back such expenditure (and regulatory infrastructure/enforcement more generally) and privatize/outsource some of these activities. There are also more covert effects, because poverty can become inter-generational (via its effects on health, education, and development of children). Again, there are historical parallels here (see above), the significant difference being that the welfare state (though weakened and under attack) moderates and hides from public scrutiny some of the starkest consequences of precarious work for the community. It is no longer recalled (at least among policymakers) that it was the same consequences of precarious work that played a critical role in the struggle for the introduction of social protection over a century ago. Indeed, in another twist of history, the neo-liberal policy push to privatize or move to NGOs areas of social security (such as employment assistance/job search aspects of unemployment support schemes) have similarities with the voluntary, localized, and mutual support type schemes whose conspicuous inadequacies helped to justify a shift to mandatory and centralized state social protection.

The growth of insecure and precarious work has also posed a series of serious challenges to regulation aimed at safeguarding the health and safety of workers and weakened unions and their capacity to defend workers—something already under attack from neo-liberal interests (Nichols, Walters, and Tasiran, 2008; Quinlan and Johnstone, 2009; Walters and Nichols, 2007). The promotion of labor-market flexibility has also undermined the effective implementation of mandated minimum labor standards (especially those related to working hours and wages)—somewhat ironically, as these laws were introduced in part to remedy the ills associated with the earlier period of precarious work (Quinlan and Sheldon, 2011). With regard to OHS legislation, the challenges posed by "flexible" work arrangements include coverage issues (including definitional issues relating to how laws define work; worker/employee, employers, and other duty holders); an increase in the number of duty holders on any given worksite, or production/service delivery task (with consequent ambiguity and risk shifting); and increases in the extent or array of risks that must be monitored by inspectors (including psychosocial risks linked to work intensification). The "new world of work," in which work has been outsourced/removed to numerous, remote, small, mobile, or transient locations (such as homes or temporary call centers), also places additional logistical chal-lenges on inspectorates. Multiple corporate legal identities (the corporate veil) can further complicate the enforcement process. Again, OHS inspectorates have not been immune to the cost-cutting pressures on government agencies under neo-liberal "efficiency" mantras or to pressures to adopt "light touch" enforcement. A growing body of research has identified and examined the difficulties that OHS regulators have in meeting these challenges as well as undermining worker representation mechanisms in OHS legislation (Bernstein, Lippel, Tucker, and Vosko, 2006; James, Johnstone, Quinlan, and Walters, 2007; Johnstone et al., 2001, 2005, 2011; Lippel, 2006; Quinlan, Johnstone, and McNamara, 2009; Walters et al., 2011).

Over the past decade, a number of governments and international agencies have acknowledged the importance of understanding how changes to work organization are impacting OHS and the limitations of existing regulatory and policy frameworks to meet this challenge. For their part, the major response of governments and OHS regulators has been to prepare guidance material and codes relating to some types of precarious work (such as temporary-agency workers) or hazards related to precarious work (such as psychosocial hazards, though even here, the focus has been on symptoms like bullying and stress rather than the underlying causes) or to conduct selective campaigns and enforcement in these areas (Lippel and Quinlan, 2011; Quinlan, 2004; Walters et al., 2011). More innovative and fundamental measures include changes to OHS laws to broaden their coverage (see Richard Johnstone's chapter in this book), expansion of coverage under minimum labor standards laws (Bernstein, 2006), supply chain regulation (Quinlan and Sokas, 2009), and an anti–social-dumping initiative in Norway. At the same time, given the global nature of precarity and the neo-liberal policies that drive it, there is a need to move beyond nation-state-based measures; and at this level, a policy/ regulatory debate has barely started. Further, the latter measures (largely the result of social mobilizations) remain exceptional, and there has been no sustained effort to question neo-liberal policies that give rise to precarity and a welter of social casualties. Only the combination of a pervasive and sustained socioeconomic crisis (the beginnings of which are seemingly evident) and massive mobilization (including the precariat itself) have any prospects of altering the present policy orthodoxies (Standing, 2011).

EFFORTS TO EXPLAIN HOW PRECARIOUS WORK DAMAGES HEALTH

Much of the research into precarious employment does not use or test a theoretical model to explain the connection between these forms of work and health outcomes. However, two existing models that have been applied extensively over the last 20 or more years to explain how work organization affects health have been applied to precarious work situations. The Demand Control Support model (DCS, also known as the job-strain model) and the Effort Reward Imbalance (ERI) model have been used to examine the health effects of downsizing, temporary employment, and other forms of contingent work (see, for example, Ferrie, Shipley, Marmot, Stansfeld, and Smith, 1998; Labbe, Moulin, Sass, Chatain, and Gerbaud, 2007; Niedhammer, Chastang, David, Barouhiel, and Barrington, 2006; Vahtera et al., 2005). The DCS model was developed by American Robert Karasek (1979). It argues that high job strain—high job demands with limited control or latitude for independent decision making—will have serious negative consequences for health and well-being. The ERI model developed by German sociologist Johannes Siegrist (1996) argues that imbalances in efforts and rewards (both intrinsic and extrinsic) will adversely affect the health and well-being of workers.

28 / SAFETY OR PROFIT?

Karasek's (1979) DCS model and Siegrist's (1996) ERI model should not be seen as separate realms of research but rather two important attempts to explain how the organization of work affects health. A number of researchers have used both models (including Marmot, Bosma, Hemingway, Brunner, and Stansfeld, 1997; Pikhart, Bobak, Peasey, Pajak, Kubinova, Malyutina, et al., 2010) and the findings have often been complementary. For example, a Finnish study of local government workers by Ala-Mursula, Vahtera, Linna, Pentti, and Kivimaki, (2005) found that employee control over their working time moderated the effects of job strain and effort/reward imbalance on sickness absence.

In a sense, both models also capture elements that are central to the capitalist mode of production, namely, the subordination of workers in terms of control of the work process and the demands made on them (in the case of the job-strain model), while effort/reward imbalance might be seen to capture some consequences of the focus on extracting a surplus from labor and the prioritizing of profit over safety.

At the same time, the application of the DCS and ERI models to workplace change has also been somewhat imbalanced, largely focused on the effects of downsizing/job insecurity on employees in large organizations and to a lesser extent, comparing health outcomes for temporary, part-time, and permanent workers. Even research into the latter is at present not sufficiently extensive or conceptually refined (between different categories of temporary or part-time workers, exposure effects arising because temporary workers may work fewer hours/have intermittent work sequences, or to consider spillover effects). For example, a number of studies have found significant gender-based differences in the health effects arising from factors such as the dual burden or double load on women arising from their care activities in the home (Gash, Mertens, and Romeau Gordo, 2006). Other studies (Schweder, 2009) also suggest there may be distinct causal trajectories for health and safety outcomes so that temporary workers might experience higher injury rates but not worse health outcomes than permanent workers.

The growth of precarious employment has also highlighted the need for the Karasek (1979) model in particular to take account of broader labor-market and social factors, a task that Karasek has begun to undertake. The model has also been seen as too task-focused, ignoring broader organizational and environmental factors (such as labor-market conditions and available social support). Johnson (2008) notes that the model was developed before Keynesian policies and an interest in workplace democratization in Europe (and Australasia) were supplanted by global neo-liberalism with its emphasis on "flexible" (including labor) markets and removing social protection, which has led to rising social inequality and power imbalances. Nonetheless, Johnson (2008: 15) concludes that the "core elements, namely, the intensification of effort, power, and collectivity continue to provide important ways of viewing the human impact of neo-liberal globalisation." Karasek responded to perceived limitations by making revisions to the model, the first step being the job DCS model, though criticisms of physiological mechanisms remain (see Shirom Toker, Berliner, and Shapira, 2008).

PRECARITY AND WORKPLACE WELL-BEING / 29

In addition to DCS and ERI, there have been several efforts to develop models that are more specifically addressed to explain how precarious employment affects OHS, namely, the employment-strain model (Clarke, Lewchuk, de Wolff, and King, 2007; Lewchuk et al., 2008) and the pressure, disorganization, and regulatory failure (PDR) model (Quinlan and Bohle, 2008; Quinlan et al., 2001; Underhill and Quinlan, 2011).

The employment-strain model takes explicit account not only of the job insecurity associated with precarious employment but also its consequences (notably, the regularity of jobs/periods of unemployment, the time and energy spent on job search, and the social support resources available to the worker). Clarke et al.'s (2007) employment-strain model (which includes the amount of time and resources spent looking for work as well as access to social support) has the advantage of seeing precariousness as the culmination of a trajectory of experiences, so that the work/life histories of workers need to be traced. In other words, precariousness is not determined solely by a worker's employment status (e.g., temporary worker) but how this intersects with labor-market and social position. For example, a temporary worker who resides in a home where the household income is over $200,000 per annum is in a very different situation than a temporary worker who must depend entirely on their own income to survive, whose work is intermittent, and has no family resources or other networks to fall back on.

Drawing on the work of Nichols (1997), Dwyer (1991), and others, the PDR model had three critical elements: economic/reward pressures, disorganization, and regulatory failure. The first factor clearly overlaps Siegrist's (1996) notion of effort/reward imbalance and Lewchuk, Clarke, and de Wolff's (2008) notion of employment strain (and the low income/ budget constraints associated with this), although it also includes the pressures of dependency/insecurity (as in a subcontracting or agency labor arrangement) as well as piecework/contingent payment systems. Disorganization encompasses the inexperience and disruption often found in short tenure work, including the weakening of induction, training, and supervisory regimes; the fracturing of formal and informal information flows among workers, as well as OHS management, including dissonance between multiple employers or different parties in a subcontracting chain; and the isolation and inability of workers to organize to protect themselves, as in the case of home workers. Regulatory failure captures the difficulty of ensuring minimum labor standards, allocating employer responsibility, and monitoring and enforcing laws in an increasingly fractured labor market, characterized by numerous diffuse or multi-employer work sites with informal or home-based work being extreme cases. It includes elements of regulatory coverage, implementation (awareness/risk shifting), and enforcement.

Table 1.1 provides a summary of key elements of the 3-factor model as well as the spillover effects discussed below.

Recent research also points to spillover effects whereby the presence of contingent workers can have adverse effects on noncontingent workers due to competition for jobs/work intensification, more fractured OHS management regimes, or the burden of removing "slack" and additional training and supervision responsibilities

Table 1.1 Key Elements of the PDR Model

Effort/reward pressures	Disorganization	Regulatory failure	Spillover effects
Insecure jobs (fear of losing job)	Short tenure, inexperience	Poor knowledge of legal rights, obligations	Extra tasks, workload shifting
Contingent, irregular payment (budgeting problems)	Poor induction, training, and supervision	Limited access to OHS, workers compensation rights	Eroded pay, security, entitlements
Long or irregular work hours	Ineffective procedures and communication	Fractured or disputed legal obligations	Eroded work quality, social interactions, and public health
Multiple jobs (works for multiple employers)	Ineffective OHSMS/inability to organize	Noncompliance and regulatory oversight (stretched resources)	Work/life conflict

(Mayhew and Quinlan, 2006). This highlights a potential problem with studies based on cross-sectional comparisons of contingent and noncontingent workers (another methodological issue is the limited number of employment-status categories considered). Other spillover effects include adverse impacts on work/product quality (which may have public health implications) and work/life conflict interactions. Other researchers have identified an important gender dimension of contingent work arrangements, spillover effects on communication, and work quality and effects on meeting family responsibilities. A Quebec study of adult education teachers (Siefert, Messing, Reil, and Chatigny, 2007: 299) found that precarious work contracts adversely affected mental health not only through job insecurity but also negative effects "on the ability to do one's job and take pride in one's work, as well as weakening the interpersonal relationships on which successful, productive work depends." Spillover effects (apart from the interaction of lack of control over working hours with work/nonwork balance; see Bohle et al., 2011; McNamara, Bohle, and Quinlan, 2011) have yet to be incorporated into the PDR model and represent the next major challenge in terms of its refinement. Further, there is growing evidence that the spillover effects of downsizing/understaffing, and use of contingent workers (such as contractors and temporary workers) can extend to the health and safety of clients and customers (Quinlan and Bohle, 2008: 496). For example, a study by Stegenga et al. (2002) found an association between nurse understaffing and nosocomial viral gastrointestinal infections in a general pediatrics ward.

The employment strain and PDR models can also be used to explain work-related injury—an area not addressed by either the DCS or ERI models (though a potential link through effects on cognitive function has begun to be explored; see Elovainio et al., 2009). Both the employment-strain and PDR models are in their early stages of development and will require rigorous testing (and possibly refinement). However, one point worth making is that both models, as well as research on precarious employment undertaken by Benach, Muntaner, Messing, Lippel, and others, points to some striking parallels with evidence pertaining to precarious work of an earlier period. These parallels include the connection between precarious work and poverty, contingent payment systems, long or irregular hours of work, and gaps in regulatory protection that are conducive to poor health among workers. Another parallel is the association between precarity and fractured and disorganized work organization— typified by elaborate subcontracting networks and remote/home-based work—that also damage health and safety. Finally, contemporary researchers are rediscovering that precarity has spillover effects onto the community in terms of impacts on children and families, public health/safety, and like. The suggestion here is not that the "new world of work" is identical to the circumstances prevailing prior to the era in which "standard employment" dominated in the mid-20th century. However, the rise of precarity and its effects on health can be better understood (and the social policy settings underpinning it challenged) if the development is placed in a broader historical context and the lessons drawn from the earlier period used to debunk neo-liberalism.

A FINAL REFLECTION

The growth of precarious work, and evidence of its consequences for worker (and community) health and well-being, provides a stark illustration of how the organization of work is central to any meaningful understanding of worker health and safety. neo-liberal policy settings have been used to reshape working arrangements without reference to the severe negative consequences this has had, and any serious effort to remedy/moderate these effects has had to rely on a substantial mobilization (by unions and community groups) and run a gauntlet of opposition (including government apparatuses supposedly controlled by progressive political elements). The growth of precarity—or flexibility in the labor market skewed in favor of those who engage workers—has exacerbated inequality at work, and this is in turn both health damaging and socially dislocating (Standing, 2011).

http://dx.doi.org/10.2190/SOPC2

CHAPTER 2

A Gender Perspective on Work, Regulation, and Their Effects on Women's Health, Safety and Well-Being

Katherine Lippel and Karen Messing

INTRODUCTION

A gender perspective on regulatory interventions in OHS requires not only a discussion of the regulation of women's work over the last century, but also reflection on gendered regulatory choices. Historically, the idea that men's health was less worthy of state protection than that of women was as much a result of gendered policymaking as was protectionist regulation limiting women's access to jobs perceived as dangerous. Today, some issues forgotten by regulators disproportionately affect women workers. This chapter begins with thoughts on gender differences in OHS regulation and then examines, through a gender lens, occupational health hazards associated with women's work and the regulatory effectiveness of current frameworks. It concludes with illustrations of ways in which the basic principles of *Safety or Profit?* (Nichols and Armstrong, 1973) play out when applied to women workers in the "modern" workplace.

THE GENDERED REGULATION OF OCCUPATIONAL HEALTH AND SAFETY: A HISTORICAL PERSPECTIVE

Women's arrival in the workplace is not a recent phenomenon. While it is true that the service sector is far larger than it used to be in developed economies and that women work mostly in the service sector and compose a significant proportion of workers in that sector, women have been active in the labor force, including in the manufacturing sector, for a very long time. Most of those killed in the New York City Triangle Shirtwaist Factory fire of 1911 were women. Early workers' compensation cases often involved women workers injured while doing manual labor. Currently, women around the world are actively engaged in industrial production, often in conditions inferior to those of men, as they are perceived as a docile and passive workforce (Mills, 2003). A very large number of women still work as domestic servants, where they are exposed to numerous hazards, and their employers, in most countries, escape regulation (Ahlgren, Olsson, and Brulin, 2012; Blackett, 2011; Hanley, Premji, Messing, and Lippel, 2010; Smith, 2011).

The first regulatory attention paid to occupational health and safety, working hours, and working conditions targeted the work of women and children in the late 19th and early 20th centuries, in a context of industrial expansion during a period when state intervention was sparse and liberalism was the dominant ideology. Acceptance of the need to regulate was gendered: states were amenable to regulating women's working conditions, but they were less so with regard to those of men. Vosko documents the tensions preceding the first international labor regulation (2010: 38–42), acknowledging resistance to regulation specific to women by some "self-proclaimed feminists," while underlining the support for protective measures by a "diverse group of women." While there was some debate as to the importance of regulating working conditions for all workers, the International Association of Labour Legislation (IALL), and subsequently the International Labour Organization (ILO), initially adopted conventions on night work and dangerous working conditions that were only applicable to women workers. The first gender-neutral recommendation, on the *Prohibition of the Use of White Phosphorous in the Manufacture of Matches*, was adopted in 1919 (Vosko, 2010: 48).

Vestiges of the early regulatory initiatives specifically targeting women's work still exist in some jurisdictions, although they were repealed in others, with the introduction of equal rights legislation prohibiting discrimination on the basis of sex/gender (European Agency for Safety and Health at Work, 2003). Women's equality rights both in North America and Europe have given rise to an increased participation of women in what is considered to be "nontraditional" employment for women, a development that, as we shall see, has raised new dilemmas with regard to women's occupational health.

Some of the reasons for regulation of women's work can be linked to women's reproductive roles: protection of the fetus sometimes required protection of women. This was not necessarily in the interest of the workers themselves. Paternalism that kept women out of night work and mines, for instance, also contributed to their exclusion from higher paying jobs. Vosko, quoting a 1932 ILO report entitled Women's Work under Labour Law: A Survey of Protection Legislation, illustrates the rationale for such legislation:

> By strictly limiting the hours of work for women, by sparing them night work, which is so exhausting and trying, and by preventing their physical organs from being deformed by carrying too heavy weights or poisoned by dangerous substances, *the legislator is really endeavouring to preserve the maternal function and to ensure the well-being of future generations* (Vosko, 2010: 47; emphasis in original).

Today, discrimination against pregnant workers is still prevalent. While in the United States, exclusion of fertile women from jobs that could be harmful to their fetus was held by the Supreme Court to be illegal and discriminatory in 1991 (Clauss, Berzon, and Bertin, 1993), more recently, in some of the *maquiladoras* in Mexico, women are forced to prove they menstruate every month to keep their jobs (Prieto, 1997).

While protecting women's reproductive functions by excluding them from the workplace did not meet with much resistance from employers, more proactive programs providing support to pregnant workers exposed to occupational hazards have been the object of considerable resistance. For example, since 1981, Québec's Occupational Health and Safety Act has required employers to reassign pregnant or lactating workers if their working conditions present a danger to the fetus, the breast-feeding baby, or to the worker because of her pregnancy. This law and others like it were initially opposed by some feminists who felt they posed a danger to gender equality (Messing, 1999). It should, however, be underlined that if the legislation has survived for over 30 years, it is because of a strong coalition between unions, women's groups, scientists, and physicians that has mobilized at least four times in response to moves to whittle down the legal protections. It is likely that the strength and success of the coalition can be attributed to the fact that women workers are seen as defending their fetuses and not themselves. It should also be noted that the implementation of this program actually helped all workers. For example, in some banks, when pregnant women were given seats, all tellers in the branch were then allowed to sit down. In some hairdressing salons, new ventilation was installed that benefited everyone.

Case law sheds light on the discomfort of expert witnesses, who must explain why exposure to dangerous substances, ordinarily considered as "acceptable" hazards for workers, becomes unacceptable because the worker is pregnant. As one expert witness put it, in testifying with regard to the worker's exposure to ionizing radiation, "the problem is that the pregnant worker has a member of the public in her womb" (Lippel, 1998: 267). For decades, Québec employer lobbies have vocally demanded repeal of this legislation, stating that reproductive health should not be seen as an employer responsibility, even if the hazards targeted by the legislation are specific to work exposures. Again in 2011-2012, this legislation was under fire and targeted for reform (Commission de la Santé et de la Sécurité au Travail, 2011).

CONTEMPORARY CONDITIONS OF WOMEN'S WORK: INVISIBLE HAZARDS AND REGULATORY CHALLENGES

Invisibility of Hazards of Women's Work

Men and women do not, for the most part, do the same jobs, and even when they have the same job titles, the actual work they do, and its effects on their health, are not the same. The gendered division of labor can be described as both "vertical" and "horizontal" (Armstrong and Armstrong, 1994). The *vertical* division of labor refers to the fact that, generally, women are found at the lower levels of occupational hierarchies, with lower pay and prestige. According to the Bureau of Labor Statistics (2009), in 2008, women working full-time in the United States earned 80% of men's average weekly salary. Women are underrepresented in management; about 37% of U.S. workers in the category of managers were women, and only 23% of chief executives were women (Bureau of Labor Statistics, 2009, Table 11). In Canada, the hourly salary of a woman is 85% of that of her male colleague, on average (Uppal, 2011). In Canada women are only 31% of managers, although they are 48% of the workforce (Ferrao and Williams, 2011).

The *horizontal* division of labor refers to segregation by profession and industry. In Canada, women are 6% of employees in construction; 19% in forestry, agriculture, and the rest of the primary sector; 30% in manufacturing; but 76% of those in clerical and administrative jobs (Ferrao and Williams, 2011). In Québec, 87% of women compared with 62% of men are in the service sector (Cloutier, Lippel, Bouliane, and Boivin, 2011: 73). Similar divisions are found in Europe (Parent-Thirion, Fernández Macias, Hurley, and Vermeylen, 2007).

Within industries, women usually work in specific types of jobs in all countries where this has been studied. Men are more often classed as manual workers (38% of men, 14% of women). In Québec, only one profession (retail salesperson) is found among the 10 most common jobs of women as well as men (Institut de la Statistique du Québec, 2010). The same study showed that 20% of women work in one of only five professions: secretary, sales clerk, cashier, preschool childcare, or office clerk. All five of these professions include a majority of women. Both women and men most often work among a majority of their own gender: 67% of those in men's top 25 professions were men, and 69% of those in women's top 25 professions were women (Institut de la Statistique du Québec, 2010). In an analysis of tendencies in employment, Asselin (2003) classified professions as "very disproportionately male" if the proportion of women in the profession was less than half their proportion in the labor force. By this criterion, she found that a substantial majority of jobs were very disproportionately assigned to one gender (221 of 506 professions were "very disproportionately male," while, by analogous criteria, 66 were "very disproportionately female").

Furthermore, even within the same job title, men and women are often assigned to different tasks and therefore exposed to different working conditions (Messing, Dumais, Courville, Seifert, and Boucher, 1994). For example, "cleaner" is one of the few jobs that are common for both women and men—12th most common for both. But observational studies showed that among hospital cleaners, women and men are assigned to very different tasks (Calvet, Riel, Couture, and Messing, 2012). A similar division of labor within job titles has been demonstrated for municipal gardeners (McDiarmid, Oliver, Ruser, and Gucer, 2000) and factory workers (Dumais, Messing, Seifert, Courville, and Vézina, 1993; Mergler, Brabant, Vézina, and Messing, 1987).

Women are disproportionately present in caregiving professions, including teaching, health care, and social work. When they work in production, they are more likely to be doing highly meticulous work like micro-electronics or very highly repetitive work requiring minutiae, like the work they do in the garment industry (Premji et al., 2008; Vézina, Tierney, and Messing, 1992). The hazards of "traditional" women's work—that done by the majority of working women—have historically been ignored by the key actors in occupational health and safety, be they regulators and inspectorates (Messing and Boutin, 1997), social security (Probst, 2009), or workers' compensation decision makers (Lippel, 1989, 1999, 2003), unions (Mills, 2003), employers (Harrison, 1996), or scientists (Messing, 1998). For example, from an OHS perspective, little attention was paid to the working conditions of garment workers, elementary school teachers, secretaries, sales clerks, and healthcare workers until quite late in the 20th century, and hazards to which they were exposed largely went undocumented. Most occupational illnesses associated with women's work were only recently included in the ILO (2010) lists of occupational diseases (musculoskeletal disorders and "mental and behaviour disorders," the latter included only in 2010), and they remain among the most controversial and contested from a workers' compensation perspective (Lippel, 2003, 2009; Lippel and Sikka, 2010).

Funding for research in OHS disproportionately targets men's jobs (Messing, 2002), both because the hazards associated with those jobs are more visible and therefore demanding of preventive action, but also because funding agencies linked to compensation boards have a vested interest in not shedding light on hazards that have yet to be documented. When asked why grant applications to investigate women's occupational health hazards were not being funded, a union representative on the advisory committee of the occupational health and safety research funding agency in Québec replied, "You don't want to prevent things that haven't even happened yet" [our translation].

A variety of reasons contribute to the relative invisibility of women's work from an OHS perspective. For example, many believed (and some still believe) that women don't really need to work. The temporary employment agency industry flourished by selling the image of the Kelly Girl, a housewife working casually in her spare time to earn pocket money (Hatton, 2011). That industry expanded to include all workers and all types of work in the services they offered, while avoiding regulation by relying on the image of the Kelly Girl: the message was that these were not "real workers" and therefore did not require legal protections that were available to workers in the standard employment relationship. Temporary employment agencies are known to expose workers to a higher prevalence of occupational injury in the United States (Smith, Silverstein, Bonauto, Adams, and Fan, 2010), Canada (Lippel, MacEachen, Saunders, Werhun, Kosny, Mansfield, et al., 2011), Australia (Underhill and Quinlan, 2011), and France (Grusenmyer, 2007), yet they maintain an aura of benign security through the image of their prototypical employee, the Kelly Girl (Hatton, 2011).

Some aspects of women's work superficially resemble unpaid household work. For example, women are disproportionately present in jobs requiring emotional labor (Hochschild, 2003). Their work in caregiving professions is often perceived as a natural extension of their role as nurturers (Grant, Amaratunga, Armstrong,

Boscoe, Pederson, and Wilson, 2004). However, a comparison of women's paid and unpaid nurturing, cooking, cleaning, and healthcare work shows important differences (Habib, Fathallah, and Messing, 2010). Paid work involves more repetition and heavier and different exposures.

One possible consequence of the similarities between some women's paid and unpaid work could be that the frontier between health issues associated with work at home and on the job may be more easily ignored. It may be harder for physicians, health and safety experts, and women themselves to identify specific workplace causes of their health problems.

When paid employment takes place in a private home, it is especially hard to make risks visible. First, it resembles work unpaid women do "for love" in their own homes, although, in fact, families will expose those who do domestic work for them to conditions more difficult than those faced by wives and mothers (Hanley et al., 2010). Contemporary policies on time off for domestic workers sometimes resemble those recommended in the 1880s:

> A mistress should be careful not to bind herself to spare her servant on a certain day in every month, as is sometimes demanded. 'Once in a month when convenient' is a better understanding (Cassell, 1880).

Second, the population is particularly isolated and defenseless, particularly since domestic servants in the higher income countries are likely to be immigrants (Ehrenreich and Hochschild, 2003; Fudge, 2011; Hanley et al., 2010).

Domestic service is not the only vulnerable job sector in which women predominate. Women, more often than men, are found in most precarious employment arrangements. They are more likely to occupy part-time employment; temporary employment, including casual work; and work for temporary employment agencies. Men are more likely to be self-employed or to do seasonal work (Cloutier, Lippel, et al., 2011; European Foundation for the Improvement of Living and Working Conditions, 2008; Fuller and Vosko, 2008). This situation contributes to the relative invisibility of the hazards of women's work because of the way statistics are kept.

For a long time, it appeared that women had considerably fewer occupational accidents than men. However, the difference becomes smaller in those jurisdictions where comparisons are based on accidents per hour worked (Cloutier, Duguay, Vézina, and Prud'homme, 2011; Messing et al., 1994) and especially when comparisons are made within the same industry or profession, in which women's injury rates tend to be higher than men's (Taiwo, Cantley, Slade, Pollack, Vegso, Fiellin, et al., 2009), possibly because of ill-adapted equipment and work situations (Messing, Seifert, and Couture, 2005).

Precariously employed workers may also be less likely to have coverage, to know their rights, or to report injury (Cox and Lippel, 2008; Quinlan et al., 2001). They thus disproportionately remain invisible to regulators, union representatives, and other traditional actors. Other studies have found that health effects associated with precarious employment differ for women and men, women being more likely to report psychological effects while men report physical effects (Kim, Khang, Muntaner, Chun, and Cho, 2008).

Dembe has shown that physicians, at least historically, did not think of their female patients as potential workers so that the documentation of working conditions associated with diseases affecting women was late in coming. Questions as to occupational exposures are less frequently asked, and the assumptions that women do not have potentially harmful work exposures have been shown to lead to erroneous conclusions that health problems more frequently experienced by women, such as carpal tunnel syndrome, are not work-related (Dembe, 1996). Stellman describes a large U.S. government study of environment and pregnancy outcome that examined pregnant women's income, living conditions, and personal habits, and asked questions about husbands' paid occupations, but did not identify the women's profession or industry (Stellman, 1978: 141). Physicians' attitudes toward women workers have been shown to be perceived as less attentive and supportive than their attitudes toward male workers and this, in turn, may have repercussions not only for acknowledgment of occupational injury but also with regard to return to work (Upmark, Borg, and Alexanderson, 2007).

Allegations of psychogenic epidemics (Olkinuora, 1984), or "mass hysteria" associated with symptoms reported by women workers, have been attributed to the failure of the medical community, and of occupational medicine in particular, to thoroughly investigate exposures of women to potentially toxic substances and the interactive effects of these exposures (Brabant, Mergler, and Messing, 1990). Some suggest that labels of mass hysteria have been attributed to women workers' actions of resistance against intolerable working conditions (Kim, 2011). The collective denial of the legitimacy of repetitive strain injury in Australia (Quintner, 1995; Reid, Ewan, and Lowy, 1991), and to a lesser extent in Canada (Kome, 1998; Lippel, 2003), illustrates the continued trivialization of women's occupational health problems. Furthermore, physicians are known to perceive women as more emotional and are more likely to resort to psychological diagnoses when women present with symptoms identical to male patients they diagnose with physical ailments (Wallen, Waitzkin, and Stoeckle, 1979).

The invisibility of women's working conditions is also an issue in occupational health research, and, for example, researchers were slow to study occupational cancers affecting women (Zahm, Pottern, Lewis, Ward, and White, 1994). However, the feminist research community mobilized around this issue, and many studies have been initiated, identifying a number of carcinogens in jobs held by women (Brophy, Keith, Watterson, Park, Gilbertson, Maticka-Tyndale, et al., 2012; Mannetje, Slater, McLean, Eng, Briar, and Douwes, 2009; Zahm and Blair, 2003).

Privatization of public services, notably in health care and social services, contributes to work intensification and the associated reduction in time available for contact with patients (Armstrong, Banerjee, Szebehely, Armstrong, Daly, and Lafrance, 2009), the breaking down of work collectives (David, Cloutier, and Ledoux, 2011; Seifert, Messing, Reil, and Chatigny, 2007), and the deterioration of worker rights (Stone, 2006). These all take a toll on worker health (Stinson, Pollack,

40 / SAFETY OR PROFIT?

and Cohen, 2005). Adverse consequences of restructuring are found in both public and private healthcare services in Canada, and it is not surprising to find that workers in the healthcare sector report the highest levels of exposure to occupational violence and various psychosocial hazards. Workers in the public sector are disproportionately exposed to occupational violence, notably in health care and teaching. In Québec, 21% of women, but only 14% of men, working in the education sector (Lippel, Vézina, Stock, and Funes, 2011), reported exposure to psychological harassment, while violence toward teachers was found to have differential consequences on men and women in a British Columbia study (Wilson, Douglas, and Lyon, 2011).

A final point in this regard: While it is true that many OHS scholars and stakeholders have historically ignored or forgotten that populations of workers include women, the same can be said for scholars and stakeholders interested in women's health who also forget women are workers. As a result, the "women's health" networks often focus on domestic violence and environmental issues, while completely ignoring working conditions to which women are exposed. It is not uncommon to find that women's committees in unions have no mandate to address occupational health and safety questions, while OHS committees have historically avoided issues specific to women's health. In Québec, collaborations between scientists and unions in studying women's occupational health have provided opportunities for these committees to learn to work together (Messing and de Grosbois, 2001).

Role of Regulatory Challenges in Understanding Gender Inequalities

At least in Québec, two issues are of importance in the understanding of gender inequalities in the context of occupational health and safety policy: the relative invisibility of women workers in occupational health and safety structures and practices, and the rejection by the powers that be of regulatory interventions regarding hazards that disproportionately affect women workers.

Currently, Québec's occupational health and safety legislation is under review because of long-standing pressure from unions to force the government to implement legislation adopted in 1979 and never brought into force with regard to over 80% of workplaces, including those sectors in which most women work. Still, in 2013, the vast majority of workplaces are not required to have either joint health and safety committees or safety representatives (Walters et al., 2011: 217–218). As a result, there are very few female health and safety militants in Québec, women rarely have access to OHS training provided to safety reps and members of committees, and women's voices are not often heard. Inspections of women's workplaces, which are disproportionately found in nonpriority sectors, are relatively rare (Messing and Boutin, 1997), and inspectors are most often trained in fields of relevance for work done disproportionately by men. The inspectorate includes few ergonomists or psychologists, more chemists and engineers.

Approximately 40% of workers are unionized in Québec, and women are now as likely to be unionized as men. However, the OHS culture within unions evolved predominantly around issues of health and safety problems affecting men, largely because men made up over 85% of the workforce in the first "priority sectors," those sectors in which health and safety committees, prevention programs, and safety representatives have been mandatory since 1980. OHS in Québec is a man's world. This affects the understanding of unions as to OHS problems of women workers, which in turn can influence regulatory developments and prevention strategies.

Regulatory frameworks promoting prevention of occupational health problems common to women's work are particularly controversial, at least in North America: regulation of biomechanical and psychosocial hazards as well as scheduling provide illustrations. Compensated injury has historically driven prevention priorities in Québec, and women's health problems are less often compensated. They are more likely to be affected by occupational diseases rather than traumatic injuries (Messing, 1998), and the regulatory appetite for preventing disease, and particularly diseases associated with poor organizational practices, is all but absent in North America, and timid even in Europe, where psychosocial (Iavicoli, Natali, Deitinger, Maria Rondinone, Ertel, Jain, et al., 2011) and biomechanical hazards (Probst, 2009) are not readily addressed by regulators and compensation authorities. While in recent studies, women were no more frequently exposed to occupational stress than men in Europe (European Agency for Safety and Health at Work, 2009), this was not always the case (European Agency for Safety and Health at Work, 2003), and they do report greater exposures than men in Québec (Vézina, Cloutier, Stock, Lippel, Fortin, Delisle, et al., 2011). Musculoskeletal disorders are prevalent in both men's and women's work, but in Québec, more common among women, especially women manual workers (Stock, Funes, Délisle, St-Vincent, Turcot, and Messing, 2011), which is also true in Europe with regard to upper extremity disorders (European Agency for Safety and Health at Work, 2003). Both musculoskeletal disorders and mental health problems have multiple etiologies, which makes it easier for authorities to externalize responsibility for their development. Furthermore, while it is important to acknowledge challenges in the regulation of chemical hazards (Jasanoff, 1987), the huge variation of ways that biomechanical and psychosocial hazards are manifested makes regulatory interventions particularly difficult to conceptualize, and even less amenable to an approach based on specifications and quantification. This is true both legally and politically.

The U.S. debacle on ergonomic regulation serves as an illustration of the political difficulties involved in developing regulatory provisions to prevent musculoskeletal disorders (Lippel and Caron, 2004). One of the most highly charged debates in American occupational health and safety regulation occurred when the federal authority, and then Washington State (Silverstein, 2007), enacted "ergonomic regulations" designed to reduce worker exposure to "ergonomic hazards." Both regulatory schemes included quantitative exposure measures and provoked a remarkably vocal backlash on the part of industry, which culminated in George W. Bush repealing the federal regulation as one of his first orders of business after taking office as president. The Washington State regulation was eventually repealed by virtue of a statewide referendum (Foley, Silverstein, Polissar, and Neradilek, 2009).

Given the relatively low profile of occupational health and safety in general, this populist anti-regulatory groundswell is indicative of the success of business in winning the hearts and minds of voters who appear, at least in the United States, to subscribe to the neo-liberal anti-regulatory discourse that suggests that regulation is antithetical to job creation and employment. Foley and colleagues (2009) have shown that while exposure of workers to hazards associated with musculo-skeletal disorders was reduced after the implementation of the Washington State regulation, since its repeal, exposure levels have increased.

In Canada, a softer confrontation has taken place, although it is unclear whether the situation there is any better for workers. Regulations on ergonomics were adopted in Saskatchewan in 1988 and in British Columbia in 1998, while other Canadian provinces have chosen to use so-called soft law to promote reduction of exposures to biomechanical hazards. While, in those provinces, without explicit regulations, general duty clauses have occasionally been applied by labor inspectorates to reduce exposures to ergonomic and psychosocial hazards (Lippel, Vézina, and Cox, 2011), the vast majority of prevention strategies are based on voluntary compliance and economic incentives. British Columbia's regulation is process based, although it is accompanied by quantitative guidelines, compliance with which is voluntary. Shortly after its adoption, a neo-liberal government came to power, which attacked all regulatory interventions, requiring organizations, including the OHS regulator, to reduce by one third the regulatory "burden" on business. Each regulation was defined as a sentence with an active verb, and several key informants reported that the B.C. OHS regulator spent a significant amount of energy parsing its regulations to reduce the number of active verbs. Regulatory provisions on ergonomics survived this period of scrutiny, as did the regulations on violence. However, cutbacks were significant during that period, leading to the reduction in the number of inspectors, and very few inspectors with training in ergonomics survived the reorganization. Because musculoskeletal injuries are very costly for workers' compensation boards in Canada, prevention is officially promoted, but in practice, enforcement of regulatory requirements, either the specific regulations or through the application of a general duty clause, is extremely rare.

Similar distaste for regulatory intervention has been expressed with regard to psychosocial hazards. Some provinces have adopted legislation requiring employers to take steps to protect workers from occupational violence and harassment, the design and strength of which vary between jurisdictions; and Ontario recently included protection from domestic violence in the workplace in its requirements (Lippel, 2011). Québec, however, is the only Canadian jurisdiction in which occupational health and safety legislation explicitly addresses work organization, key to the reduction of psychosocial hazards, as part of the employer's general duty clause governing prevention. This has allowed for occasional interventions by the inspectorate and the appeal tribunal with regard to staffing and implementation of unhealthy technological innovations (Lippel et al., 2011), but nothing comparable to the potential of regulation in countries like Australia, where risk assessment during periods of restructuring is required (Quinlan, 2007). Ironically, Québec legislation on psychological harassment was included not in the occupational health and safety framework, but in general legislation governing labor standards (Lippel, 2011). Traditional OHS actors, including workplace parties and the regulator, showed little enthusiasm for regulatory action on harassment, demanded by NGOs representing nonunionized workers (Lippel, 2005). This explains the inclusion of the provisions requiring the employer to prevent psychological harassment and compensate targets in the Labour Standards Act, although these requirements have now stimulated unions to take a proactive stance in the prevention of harassment.

Even when legislation explicitly addresses workplace violence, as in British Columbia (Sections 4.24–4.31 of the Worksafe BC Occupational Health and Safety regulation), the legal language used may be seen to obfuscate the importance of worker-on-worker violence, an issue of paramount importance for women, particularly when they are in a minority in the workplace, occupying nontraditional employment. Thus, in British Columbia, worker-on-worker violence is excluded from the definition of violence in the OHS regulations, and is labeled "improper activity or behaviour," defined as

> the attempted or actual exercise by a worker towards another worker of any physical force so as to cause injury, and includes any threatening statement or behaviour which gives the worker reasonable cause to believe he or she is at risk of injury; and horseplay, practical jokes, unnecessary running or jumping or similar conduct.

Sexual assault by co-workers would be included in this definition. Not only does the language trivialize the activity, but the requirements of employers are softened with regard to "improper activity or behaviour," and no risk assessment is required (Lippel, 2011). Prevention of psychosocial hazards is targeted by soft law in Canada as part of a nonregulatory standard leading to, at best, voluntary compliance by industry (National Standard of Canada, 2013). The fact that workers' compensation is not provided to workers who develop mental health problems attributable to poor work organization in most Canadian provinces (Lippel and Sikka, 2010) suggests that the economic argument for reduction of exposures is muted in those provinces where disability attributable to workplace psychosocial hazards is noncompensable.

Regulation of work schedules and precarious employment, notably that provided by temporary work agencies, is known to be of primary importance for the protection of workers' mental and physical health, yet suggestions as to the necessity of regulation on these issues meets with considerable hostility, both from the temporary employment agencies themselves and also from all employers, including the state, who see scheduling and externalization as the prerogative of the employer.

Because of their greater responsibility for caregiving and domestic work, women are disproportionately affected by a deregulated labor market in which anything goes in relation to work schedules. When workplace demands are increasingly allowed to invade the home sphere, as with unpredictable and variable scheduling, women suffer disproportionately. So far, workplace conditions that render work/family balancing difficult have been infrequently considered in labor legislation and enforcement. A 2-day notice of hours to be worked in the following week or mandatory evening shifts and weekend shifts are becoming increasingly common in the service sector. In Québec, as elsewhere, women are more likely to have a heavy domestic workload and yet are less likely than men to have access to working conditions that facilitate work/life balance, including rest breaks or the possibility of dealing with personal matters at work (Cloutier, Lippel, et al., 2011, annex Table 34; Lippel, Messing, Vézina, and Prud'homme, 2011, Table C3.13).

Work/life balance was recently studied in a large retail sales chain in Québec, whose work schedules were generated by software that based its output on predicted sales per 15-minute period. The resulting schedules were extremely variable and unpredictable and were posted with little notice: on Thursday or Friday for the week beginning the following Sunday or Monday. Shifts could last anywhere from 4 to 8 hours and could start any time between 5 a.m. and 7 p.m. Weekly holidays could be during the week or on weekends and could be on separate days or back to back; 83% of workers had been assigned to work during the weekend preceding the survey. Pay was low, near minimum wage levels. As a result, few workers had any kind of family responsibility; only 14% had a child under 18, about one-third of the rate for other Québec workers (Institut de la Statistique du Québec, 2008a, 2008b). One-third of all workers reported that conditions for work/life balance were "difficult": women, especially those with family responsibilities, were significantly more likely to report difficulties. The presence or absence of family responsibilities did not affect men's reports of difficulty in balancing work and personal life (Messing, Couture, Tissot, Bernstein, Lienart, Rousseau, et al., 2011).

Women are also less likely than men to report decision authority at work and more likely to have high demand at work; they have less control of work content and work speed (Vézina, Stock, Funes, Delisle, St-Vincent, Turcot, et al., 2011, Table 4.3). All workers, but particularly women, would benefit from statutory protections giving them more control over their schedules, yet given the neo-liberal environment, there is thus far no indication of upcoming regulatory attention to this issue.

The fact that the acknowledgment of the need for regulatory intervention to protect workers from hazards that are having a significant impact on women's health has come of age in an antiregulatory environment certainly contributes to the sluggishness of regulatory bodies, which are currently reticent to strengthen protections against universally acknowledged lethal hazards like silica (Greenfieldboyce, 2012).

AND WHAT OF THE GENDER PERSPECTIVE ON SAFETY OR PROFIT?

In the same way the Robens Report (1972) alleged that apathy was the cause of occupational accidents (Nichols and Armstrong, 1973), women workers have been, and continue to be, accused of irresponsibility and failure to protect themselves from occupational hazards. In 2012, production still trumps safety. However, production in women's work is often driven in more subtle ways. While production quotas and pressures to maintain work at a pace incompatible with occupational health and safety remain common in the manufacturing sector (Premji et al., 2008), most "production quotas" in women's work are more subtly expressed—in the obligation to provide care to patients, to teach children, and to serve patrons without the staffing necessary to provide adequate service. Women forced to choose between protecting their own health or that of those they serve will be inclined to place the patients and the children ahead of their own OHS interests, and will be perceived as imposing on themselves working conditions that have become untenable and intrinsic to public service in an era of cutbacks in both the private and public sectors.

Three decisions drawn from workers' compensation case law in Québec are reminiscent of the examples provided in Nichols and Armstrong's *Safety or Profit* in 1973.

In one of the earliest claims for workers' compensation for mental health problems attributable to work-related stress in Québec, a workers' compensation board team leader filed a claimed for a depression related to the conditions in which new workers' compensation legislation was implemented. She described how legislative reform affecting hundreds of thousands of claimants was put into force prior to training of frontline staff, who were inundated with requests for explanations without having been provided with answers or strategies to deal with the incoming calls. As the team leader, she answered the phones that the frontline staff refused to answer, doing her best to fill in for the system that was failing the injured workers. She fulfilled this role for months and was eventually diagnosed with "burn out or situational depression." Her workers' compensation claim was denied for two reasons. First, the judge found the situation was not sufficiently abnormal to meet legal compensation requirements, and second, she was described as having a pathological attitude toward work, which led her to fill in for the system when the system failed. For the employer, its expert witness, and the judge, it was her predisposition to perfectionism that was the cause of her illness. This case was decided in 1987.[1] Since then, there have been thousands of claims for workers' compensation for mental health problems related to work in Québec; some accepted, most denied. Often they are contested by employers or denied by the compensation board because the mental health problem is judged to be the workers' fault for caring too much, for putting their own health on the line to protect the health of others. Employers value perfectionist employees, as they are more meticulous and productive. The worker's perfectionism is then turned against them if they become ill because of their devotion to their job.

Case law also illustrates ways in which employers use workers' conscientiousness and commitment to the job as an argument to contest their compensation claims. This

---

[1]*Laflamme et CSST* (1987). CALP 00725-60-8608, December 4, 1987, AZ-4000000520. See also *Barrett et CSST*, 2007 QCCLP 4953.

strategy ultimately failed in a case in which a nurse in the public sector suffered from depression after having spent 6 years adapting to continual restructuring that required her to provide care in 7 different types of specialty clinics, as well as train subordinates when she herself had not had adequate training. Her depression was triggered when she was involved in the administration of inappropriate medication that jeopardized the health of a child, and she told the mother of the mistake and her institution's potential liability. Medical evidence referred to "adaptation overdose," but other medical experts attributed her depression to her "obsessive compulsive" personality and her inability to deal with the moral dilemma raised by sharing the information with the mother of the child, as she felt she had betrayed her colleagues and her employer. The appeal tribunal eventually accepted her claim, concluding that the depression she had suffered was an occupational disease, as the continued changes and adaptations required of her were occupational hazards responsible for her condition.[2] Yet the message sent by the employer who contested the claim, and by the compensation board that had initially denied it, strengthened the belief that those who suffer injury shall be blamed and punished if they dare to speak the truth about the process that led to their disability.

Safety or Profit showed how regulators turn a blind eye to the reality of "a society in which some men are paid to squeeze as much production as possible out of others" (Nichols and Armstrong, 1973: 30), and illustrated how "men exert such pressure upon themselves and their workmates in anticipation of the demands they believe management would make upon them if they failed to do so" (Nichols and Armstrong, 1973: 19). If anything, this process has been exacerbated in current times, and women, like men, are subjected to work organization designed to squeeze out the maximum amount of work by encouraging seemingly self-imposed pressures to increase productivity. In a recent illustration, the compensation board initially denied the claim of a woman working in the retail service sector. In the *Tardif* case[3] restructuring had forced the worker to choose between layoff and occupying a new position at a reduced salary. Economic incentives to work longer hours were intrinsic to her new contract, and she worked more than 50 hours a week for eighteen months, until she fell ill. Diagnosed with depression, she filed a workers' compensation claim that was denied because the compensation board concluded her overwork was self-imposed, attributable to her own negligence and personality. The appeal tribunal accepted her claim, concluding that workers' compensation in Québec is a no-fault system, and that blaming the worker, in the absence of gross negligence, did not justify denial of the claim. The appeal tribunal also took note of the fact that the employer benefited from the claimant's overcommitment.

In these examples, some workers did receive workers' compensation benefits, which in part is why we know about them. But thousands of workers see their claims

---

[2] *Plouffe-Leblanc et CHUS Hôpital Fleurimont* (2003). C.L.P.E. 2003LP-97. (Plouffe-Leblanc, 2003).

[3] *Tardif et Repentigny Mitsubishi et CSST,* 2011, QCCLP 4744.

denied, and thousands more never apply for compensation, deeming their health problems to be noncompensable and sometimes, perhaps not related to work, the boundaries between caring at work and caring at home being sufficiently muddled to lead workers to believe their pain and their problems to be exclusively personal.

For women, safety or profit is not the only dilemma that affects their health at work. Women are also forced to weigh safety against equality considerations. When they enter nontraditional occupations, for example, women who complain about health risks can be seen as unfit for the job (Messing and Elabidi, 2003). Even in more "feminine" professions, women have to contend with the suspicion that they are fearful, complainers, or weaklings. The idea of communicating any hint that women suffer gender-specific health effects can trigger fears of discrimination and lead women not to use available recourse (Malenfant, 2009).

CONCLUSION

Nichols and Armstrong's *Safety or Profit* (1973) denounced as illusory the assumption underpinning the Robens Report (1972), that workers and management have a shared interest in health and safety. The contemporary dominant OHS discourse remains the same, and the critique is just as relevant today and perhaps even more so with regard to women workers. In addition, continuing gender stereotyping and discrimination against women workers counters efforts to gain recognition and compensation for women's occupational health problems.

On many levels, women and men are exposed to similar organizational hazards that have become worse since the publication of *Safety or Profit* (1973): work intensification (European Agency for Safety and Health at Work, 2007), job insecurity (Quinlan and Bohle, 2009) that drives hyperwork (Carpentier-Roy and Vézina, 2000) and presenteeism (Dew, 2011), and new organizational strategies to make workers collectively responsible for team productivity (Roy, 2003), and thus collectively responsible for poor OHS choices that superficially appear to be self-imposed. Yet, in other ways, women workers encounter additional dilemmas to those identified in *Safety or Profit*.

Because women are more likely to do work involving care for others, and because those institutions that employ them are under increasing pressure to cut services and slash costs, women are disproportionately exposed to psychosocial hazards and then labeled as having personality disorders when they become ill. The war on public services in Canada as in Britain, disproportionately affects jobs done by women. Cutbacks in male-dominated sectors like manufacturing lead to externalization, plant closings, and massive unemployment; cutbacks in the public sector, while also leading to layoffs and shutdowns, leave some workers behind to do more with less. When they eventually fail to meet impossible objectives, they blame themselves for the system failure and for their own perceived failures. And were they to become ill, they are told (and often believe) that their problems are attributable to personal failings, either because they fail to cope with their impossible working conditions or because they have personal problems or personalities that explain the development of their illness. Furthermore, providing care is often perceived to come naturally to women, both in paid and unpaid work. It is not difficult for regulators and employers to believe and to convince women that their poor health is attributable to personal failings or problems at home rather than to process failure in the organization of their work. The reduction in public sector employment, and the fact that public services are increasingly in competition with the private sector, also contributes to a downward harmonization of contractual and regulatory protections (Stinson et al., 2005).

Dominant discourse on the "less law" approach, discussed in *Safety or Profit* (1973), is stronger than ever in 2012. Regulatory protections that could be of use in the prevention of hazards to which women are disproportionately exposed are either nonexistent or sufficiently nonspecific to defy enforcement. Those regulations that do provide tangible protections to women workers in Québec, like the provisions on protective reassignment of pregnant or lactating workers, have recently been the subject of heated debate in the context of the "Modernisation" of Québec's occupational health and safety legislation (Commission de la Santé et de la Sécurité au Travail, 2011). While union and industry representatives approved, in principle, a reform that would most likely transfer ultimate control of the program from the hands of the treating physician to the hands of the regulators, women workers and the groups that represent them vociferously resisted these proposed changes. Industry says that costs are out of control for this program. Rather than eliminate the dangerous exposures, which would cost money, employers prefer to reframe the discussion of a program designed to provide protection from hazards in the workplace that differentially affect pregnant women and their fetuses, labeling it as a "social" program whose cost should be underwritten by the state and not employers, an attempt to justify the whittling away of the program itself.

In 2013, costs, including costs of workers' compensation for occupational injury and disease, are the ultimate drivers of prevention interventions, including regulatory interventions. Current offensives to reduce costs of compensation by indiscriminately promoting early return to work (MacEachen, Ferrier, and Chambers, 2007; MacEachen, Kosny, Ferrier, Lippel, Neilson, France, et al., 2011) and attempting to shut out claims for musculoskeletal disorders and mental health problems will, if successful, allow for business as usual: putting profit before safety, for women and men alike.

PART II

New Governance, Organized Labor, Deregulation, Decriminalization, and the Neo-liberal Agenda

It is rarely possible to consider issues that bear on health and safety at work without reference to Sweden. In Chapter 3, Kaj Frick examines the Swedish work environment system and in particular the position and influence of safety representatives as they have coped with the advent of a center-right government, higher unemployment, and an increasingly divisive labor market. In Sweden, as elsewhere, state and employer practice has been inflected by the rise of so-called new governance theory. This forms the subject of Chapter 4 in which Eric Tucker explores, largely but not exclusively in a Canadian context, just how new "new governance" is. He traces its roots back to the Robens Report (1972) and makes clear its affinity with neo-liberal self-regulation. In Chapter 5, attention switches to the UK, the site of the original Robens Report. Here, Tombs and Whyte document the acceleration into the neo-liberal turn that took place under New Labour. They point to the coming to prominence of the "business case" for health and safety, the rise of the "Better Regulation" agenda, and the collapse of regulatory intervention, workplace inspection, enforcement notices, and prosecutions. In Chapter 6, Richard Johstone reviews the several Australian state jurisdictions and argues that, despite instances of more robust prosecution, there is continuing ambivalence about whether health and safety at work cases are "really" criminal and that, in practice, the legal structure surrounding prosecution contributes to decriminaliszing work, health, and safety offenses.

http://dx.doi.org/10.2190/SOPC3

CHAPTER 3

Resilience Within a Weaker Work Environment System—The Position and Influence of Swedish Safety Representatives

Kaj Frick

DEVELOPMENT AND EROSION OF THE SWEDISH WORK ENVIRONMENT SYSTEM

Voluntaristic and Cooperative Work Environment Traditions

This chapter describes and discusses the situation of Swedish safety representatives (SRs) with respect to formal rights and position, power balance and influence, and the social construction of risks and solutions. It is prefaced by a short account of the development of the Swedish work environment (WE) system and how recent economic and political changes may have affected this and the position of safety reps. An underlying question is, Why do safety reps refrain from fully using their rights and opportunities to act against risk at work? They may pursue any serious risk as much as is needed. And work environment surveys indicate widespread and serious risks at work that the SRs (and others) could require employers to improve (AV, 2010). The risks are estimated to cause at least 1,000 fatalities per year, but possibly many more (Järvholm, 2010). Despite this, in 2006, only 3% of the blue collar SRs in the Swedish Trade Union Confederation (LO) survey had used their right to appeal to the labor inspectorate during the previous 3 years (Gellerstedt, 2007: 49–50). Case studies also indicate that safety reps do not use all of their rights to pursue issues they find important (e.g., Frick, 1994; Frick and Forsberg, 2010). To at least somewhat understand the difference between theoretical and real safety rep influence, we have to look not only to their formal position but also to their labor-market power and how risks are constructed and understood at the workplace; and how these factors have changed in the Swedish labor market.

Swedish safety reps operate within a mainly cooperative and consensus-oriented tradition. From the start in the late 19th century, the state's work environment (then "safety") politics tried to achieve voluntary reductions of occupational risks through advice and persuasion. Although there are now many and strict WE regulations, compliance is still mainly to be achieved through local cooperation between the social partners. In its present form, the work environment system evolved through a number of legal and other reforms during the 1970s, including the large Work Environment Fund of 1972 (to promote R&D, information and training), and the Work Environment Act of 1978 (Frick, 2011a). This act requires a high WE standard of employers. The Swedish Work Environment Authority (SWEA) issues, supervises, and enforces provisions that specify the legal requirements, albeit with few details on psychosocial risks (Bruhn and Frick, 2011). SWEA's staff and regulations grew after the 1970s reforms. Despite later funding cuts, the authority still inspects some 5% of all workplaces per year (AV, 2011a). Yet formal enforcement or sanctions are rare, both against violations of the provisions and after accidents at work. Employers mostly comply with SWEA's nonbinding inspection notices without the need for formal enforcement actions (Frick, 2011b).

This voluntarism was inspired by the U.S. Safety First Movement (Lundh and Gunnarsson, 1987) and by Swedish consensus traditions (Rothstein, 1992). After decades of industrial conflict, the social partners reached a central agreement on mutual acceptance and cooperation in 1938. They took charge of safety/work environment politics from the 1940s, partly as a strategy for the employers to minimize state intervention (Steinberg, 2004). There were central, industry-level and local agreements between employers and unions on WE cooperation. These supported the local dialogue and joint improvement projects and programs. The state supported this cooperation by promoting worker participation. With the strong position of Swedish unions, the participation is mainly through safety representatives. In the inquiry behind the Work Environment Act (SOU, 1972: 117–132), the social partners jointly supported this system with arguments about how reps informed managers of risks and co-workers of the importance of following safety rules.

For many years, Sweden has also given considerable state funding to an "enlightenment" strategy to improve voluntarily through advice and persuasion. Research defines risks at work and their causes, while development projects find solutions. Information and training spread knowledge of this to the workplaces. This was largely done by the social partners and (until 1993) by subsidized advice from multidisciplinary occupational health (OH) services that used to cover some 80% of all employees (Frick, Eriksson, and Westerholm, 2005). The 40-hour basic WE training, developed jointly by the social partners, reached very large groups of SRs and their supervisors in the late 1970s (Kamienski, 1979). Local actors do not use all available knowledge on what to do, but such information is at least easy to find if they try to manage occupational risks. The extensive enlightenment has also spread a social norm that risks at work are serious, that responsible employers/managers can and should prevent them, and that this prevention often is cheap (Frick, 2009). Through all of this, the enlightenment supports the normative power of SRs. Applied to Denmark, Dyreborg (2011) discusses this local dialogue and cooperation (with strong worker/union involvement) as the democratic work environment regulatory strategy.

Voluntarism and the safety representatives' influence within local cooperation is also significantly based on the strength of labor. Sweden's economy, labor market, and politics long supported this. Workers' unions were strong, and capital had to learn to cooperate with them. This included combining improvements in productivity and work environment (Frick, 2011a). Safety reps (and other union stewards) had instrumental economic power to influence managers mainly because

- Sweden had a strong economic growth from 1870 to 1970. This gave capital enough profit to share some with labor (and the government through taxes), also as WE improvements.
- The growth created a strong labor market. Unemployment was rarely above 3% from 1945 to 1991. Employers had to improve jobs to recruit, retain, and motivate workers. They had to listen to the SRs as the worker voice and to reduce the very costly exits of dissatisfied workers (Frick, 1994; Hirschman, 1970).
- Union density rose to 83% in 1993 (Kjellberg, 2011). Unions often had the ear of labor governments (during 1932–1976, 1982–1991, and 1994–2006).

Weaker Labor and Work Environment Systems After the Early 1990s

The conditions that supported labor, safety reps, and local WE cooperation peaked during the early 1970s, with a good economy, strong labor market, high unionization, and a prolabor government. When the 1978 Work Environment Act was to be implemented, the situation had already started to deteriorate (Kronlund, Fleischauer, and Grünbaum, 1985). However, the severe economic crisis of the early 1990s much accelerated changes in the economy and in politics on the labor market and its work environment (Frick, Eriksson, and Westerholm, 2009, 2011a; Wikman, 2010). The major changes were the following:

- The employer organizations terminated nearly all WE agreements in 1992 as being neither necessary nor adapted to the varying conditions of individual firms. These have been replaced with fewer and usually more general agreements. The national- and industry-level cooperation also continues, but less than in the 1970s and 1980s.
- The OH services were deregulated in 1992. The services now give much less preventive advice and support to managers and safety reps than they did before 1992 (Frick, Eriksson, and Westerholm, 2005).
- Since the recovery later in the 1990s, the Swedish economy (and public budgets) have done relatively well internationally. However, half of the GDP is made up of exports and imports. Business and employment is very dependent on the global economy. There has also been a large job export, especially in

manufacturing. Unemployment has increased. An unemployment ceiling used to be 4%; now it is a minimum floor to strive for.

- The economy is transforming from production in hierarchical organizations toward market production in supply chains. Labor hire, outsourcing, small firms, self-employment, franchising, casual employment, and subcontracting are increasing. This is promoted by the present government, which has eased several labor regulations, including for-labor hire and temporary employment. Former public activities are being privatized, mainly education plus health and other care, though most of this is still produced in the public sector, which presently represents 31% of all employment.

- The economy has also shifted from blue collar work in manufacturing toward white collar jobs and also less skilled service jobs. The growing private services have smaller firms with a faster turnover and more casual jobs than the other two major sectors of manufacturing and public services.

- More international ownership has resulted in less CEO support for the Swedish model of local cooperation between managers and unions (Levinsson, 2004).

- The new center-right government, which came to power in 2006, cut SWEA's funding by a third from 2006 to 2009. All of SWEA's activities were reduced, including some 20%–25% fewer workplace inspections. Although SWEA still has to respond to formal SR requests for inspection, they provide less support to the reps and other local WE actors.

- Posted and other migrant workers, especially after the EU enlargement in 2004, have increased unfair business and labor-market competition, notably in construction, transport, cleaning, and restaurants. The EU court has circumscribed union rights to use industrial action to uphold collective agreements (and indirectly also the WE standard) for these workers. At the same time, SWEA has fewer inspectors to supervise the many new and often short-lived migrant workplaces (Frick, 2009).

- After the 1990s crisis, the WE enlightenment got gradually less public funding for R&D and for spreading its results. This decline accelerated in 2007 when the government closed the National Institute for Working Life and withdrew subsidies for training and information.

- The new government also reduced eligibility and compensation for unemployment, worker compensation, and sickness insurance. It has become much more costly not to be at work. The government also increased membership fees to unions and to the unemployment insurance run by them. This accelerated the deunionization from 83% density in 1993 to 70% in 2011. White collar unions have retained more members (now 73%) than the blue collar ones (67%). The difference is larger between the public sector (with 80%) and the private service sector (with 55% union membership; Kjellberg, 2011; LO-tidningen, 2012). However, the largest gap is between the permanently and the temporarily (14% of all) employed, with only some 40% of young temporary workers being organized (LO, 2011).

THE SAFETY REPS' POSITION, POWER BASE, AND SOCIAL RISK CONSTRUCTION

Earlier Reports on Safety Reps

So where have the changing economic and political conditions left Swedish safety representatives? To understand the safety reps' present situation, it is useful to briefly draft how this was when the reforms had just started and also when Sweden was recovering after its economic crisis. In 1981, Keisu Lennerlöf described the role of safety reps within SACO unions (the Swedish Confederation of Professional Associations, i.e., employees with academic qualifications). She also commented on the reps in the blue collar unions in LO and the general white collar unions in TCO (the Swedish Confederation of Professional Employees). The work environment reforms were new in 1981, and many more safety reps had been appointed in all three federations. Keisu Lennerlöf reported some 2,300 reps in the SACO unions (or 1 per 100 members), 23,000 in TCO unions (1 in around 45), and 82,000 SRs in the LO unions (1 rep in 20–25 members). TCO and SACO had developed work environment policies during the 1970s, especially with the broadened WE concept. They and several of their unions had initiated research on risks in white collar work, done members surveys, adopted programs, and started to train their safety reps. However, SACO reps still reported problems with a diffuse role as white collar SRs and that psychosocial risks were hard to grasp and not prioritized by ignorant managers. They often faced too little interest and support from their members and sometimes from their local unions. This could make it hard to recruit new reps (Keisu Lennerlöf, 1981). At the same time, the 1980 survey of LO's safety reps indicated a more established role and cooperation with management, with member and union support but sometimes with difficulties about resolving technical risks (Bolinder, Magnusson, Nilsson, and Rehn, 1982).

Many economic and political changes accelerated during and after the economic crisis of the early 1990s. ATK's Arbet report (2000) on SRs' need for training and information is therefore a midterm evaluation of how these changes have affected the safety reps. The main findings were that the safety reps:

- Had been reduced to some 85,000. Some 30% of the single safety reps (only one in the workplace) had no training, but in larger workplaces, around 90% of the reps had been trained.
- Faced increasing psychosocial risks from stress caused by rationalization and also from threats, violence, and harassment. Technical risks were somewhat reduced, mainly by changes in technology, but remained serious hazards. Psychosocial risks were still hard to define, handle, and often to get management interested in to assess and abate.
- New agreements to integrate the local dialogue on codetermination and on WE issues had often resulted in work risks getting less attention than before. The integration within the local unions was estimated to be slightly better, but safety reps were sometimes kept out of major union issues.

56 / SAFETY OR PROFIT?

- The reps often lacked time, with a set (but insufficient) time for their tasks, pressure from peers and/or one's own job not to "waste" time as a rep, and management not organizing a substitute to do the rep's job when s/he was to act as rep.
- The situation for SRs was assessed in a qualitative survey to all unions. In general, the unions found that the reps' position had deteriorated, with less participation in WE management, less time for training and for other aspects of their tasks, and less information from managers (though with the Internet, media information on risks at work had increased). Yet the unions still found existing reps to be active to improve WE conditions.
- The unions found it increasingly hard to recruit new and replace old safety reps, because of a combination of indifference among younger members, individual salaries for blue collar workers (with a fear that employers would not grant reps any pay raises), pressure from one's job and peers not to take time for such extra tasks, and general fear of being victimized.

The journal *Arbetarskydd* (2000) had a special appendix on safety reps, partly based on the ATK (2000) report but also on its own material. It found that the ratio of safety reps had declined from 279 in 10,000 employees in 1993 to 212 in the year 2000. Safety rep training had also been reduced, with both a shortened basic training from 5 to 3 days and more reps without any training at all. This may have partly been caused by employer resistance but much more by individualized tasks also for blue collar workers with more personal responsibility. Many workers were therefore more focused on their personal career and less interested in common union activities. White collar unions found it extra hard to recruit (and thus had lost more) safety reps in small workplaces. But the increasing work pressure also made it hard to take on extra tasks as safety reps or other union activities in larger workplaces. To do so might obstruct one's career and professional development. Several union officers who were interviewed reported that the tougher safety rep climate was an effect of less WE training of managers. Later *Arbetarskydd* journals (2003, 2004) reported that the number of reps had recovered to some 103,000 in all, with around 5,000 in SACO, 30,000 in TCO, and 68,000 in LO. But there was still only 1 rep for 117 SACO members against 1 in 42 TCO members and 1 in 28 LO members. The rep shortage remained acute in private services, also in LO's blue collar unions there. SACO's highly qualified members still found a conflict between promoting their jobs and careers and being a safety rep, along with other union roles.

Formal Position: Rights, Coverage, and Time for Their Task

Safety Reps' Rights and Roles

Safety representatives are appointed by local union organizations (or by the employees in nonunionized workplaces, but this is rare). If there are more reps in a workplace, one is to be appointed senior safety rep to coordinate all SRs. Unlike other union representatives, the SRs are to support and speak for all employees in the area, team, department, and so on for which they are appointed, not only for their union members. The Work Environment Act (ch. 6) gives the reps strong rights to participate and to engage management in an active and competent dialogue on anything that may affect health at work (Steinberg, 2004). SRs have a right to

- Be informed of and participate in the planning of all work environment changes and in the mandatory systematic WE management;
- Get training and information, see documents, and get access to premises;
- Take the paid time needed, e.g., to investigate issues and to talk to employees; and
- Carry out their tasks without being obstructed, discriminated against, or harassed in any way.

By raising and arguing for their requests—often with reference to WE provisions—safety reps can put employers (i.e., managers) under pressure to act. They must rapidly answer written improvement requests from the SRs, or the rep may otherwise ask SWEA to directly inspect the workplace. However, the safety reps rarely bring issues to open conflicts. SWEA only had to inspect and mediate in some 300–400 such cases per year during 2006–2010, most often in the public and in the transport sectors (Du and Jobbet, 2011). When reps find risks to be acute and serious, they may make the even more conflicting decision to stop work, also pending a verdict by SWEA. From 2004 to 2006, SWEA resolved some 100 local stops per year by LO reps (Gellerstedt, 2007: 65). SWEA involves the SRs in the authority's inter-actions with the employers but gives them no other support in case of conflict with the employers (Frick, 2011b). Reps should instead get union support, if needed, to sue the employer for damages for violating their rights.

Two recent changes have somewhat adapted the SR system to the supply chain economy. There is a stronger coordination (and planning) duty on construction sites (adapting to the EU directive on this). Safety reps should thereby be able to have dialogue with someone in charge at these sites, which often have multiple subcontractors. Reps may now also speak for agency labor, stop their work, and/or appeal to SWEA on behalf of them. Likewise, employers have to include the prevention of risks to agency workers in their work environment management. However, the government has not enacted other proposals (SOU, 2007, 2009) to give purchasers and main contractors some duties for the work environment within their suppliers/subcontractors activities, to partly offset their lack of responsibility in supply chains.

Since 1993, employers have had to organize and conduct Systematic Work Environment Management (SWEM; AFS, 2001). The SWEM provisions imple-ment the EU's Framework Directive (89/391/EEC). Although employers far from fully comply with the requirements, their WE management has nevertheless improved. Risk assessments and action plans are normal, at least in larger work-places, though their coverage and implementation may be imperfect (Frick, 2011b). The role of safety reps has been affected by the SWEM-regulation. Especially in large

58 / SAFETY OR PROFIT?

and highly unionized workplaces, managers had often taken a role of benevolent passivity. Daily WE activities were largely left to the reps, who informally fixed easier problems and asked for (and often got) funding for larger investments (Frick, 1994). When managers try to do what they should, it may be difficult for active reps to change into a more passive, supervisory role in how the work environment is managed (Johansson, 2010a).

The SRs' activities are also often regulated by work environment agreements. Despite the shortage of national/industry-level agreements, 62% of the LO safety representatives reported in 2006 that there was a local agreement at their workplace. WE policies and procedures were their most common content (53%; Gellerstedt, 2007). However, there are also some detailed industry agreements mandating, for example, the amount of work environment training for reps and managers (Livs, 2011). Especially within the public sector, many agreements aim to integrate the dialogue on codetermination and on work environment issues. In practice, the emphasis may be more on getting acceptance from the local union in the former than on the employer duty to handle risks at work. The integration may therefore not necessarily support an active safety rep role (Frick and Forsberg, 2010).

Coverage by Sector and by Workplace Size

Few safety representatives were appointed at first after workers got this right in 1912, with SRs in only 861 workplaces in 1930 (Lundh and Gunnarsson, 1987: 82). However, growing unions and stronger SR rights, the central agreement in 1942 on occupational safety cooperation between employers and unions, and finally the legal and other WE reforms of the 1970s resulted in more and more safety reps, with a maximum of some 115,000 in 1990–1992. After a dip to 85,000 in 2000, there are again some 101,000 safety reps (Du and Jobbet, 2012; Gellerstedt, 2012; Magnusson, 2012). LO's shrinking blue collar unions still have 64,500 safety reps, or an average of 20 members per SR, but this ratio varies much by industry. LO unions in the traditionally safety-oriented industries of mines, manufacturing, and construction and in the highly unionized public sector have 18, while the less organized and safety-oriented private services have 33 members per safety rep. In 2004, there were reps in only 3% to 10% of private service workplaces compared with 21% of construction sites, 35% of factories, and a maximum of 59% of workplaces with SRs in public administration (Arbetarskydd, 2004).

In 2004, safety reps were appointed in less (or much less) than half of the 120,000 workplaces with at least five employees, where they should be (Frick, Eriksson, and Westerholm, 2005). As SRs are rare in small firms, 70% of the LO reps in the 2006 survey were in workplaces large enough to have joint WE committees. This is an increase from the 57% in the 1996 and the 47% of SRs in workplaces with such committees in the 1980 surveys (Bolinder, Magnusson, Nelsson, and Rehn, 1982; LO, 1997). Even 21% of single SRs (the only rep at their workplace) had such committees in 2006 (Gellerstedt, 2007). To compensate for the lack of SRs in small firms (and to support 12,000 single reps in small firms), there is a system of

RESILIENCE WITHIN A WEAKER WORK ENVIRONMENT SYSTEM / 59

union-appointed regional safety reps (RSRs). The RSRs now support workers and WE cooperation in the large majority of small and microworkplaces (Frick, 2009). However, even with RSRs included, the safety rep rate per employee in small workplaces is only about a third of that in the larger ones (Frick et al., 2005).

Despite fewer blue collar workers, a lower unionization rate and a fracturing of the labor market into smaller firms and more casual labor, the LO unions have largely upheld the number of reps. IF Metall, the major manufacturing union, lost very many jobs during 2008–2009, but not many SRs. The reps were often retained when jobs were cut. The union has also appointed new SRs and prioritized support for them. As a result, their 14,000 reps means one SR for every 19 members. This means that there are SRs at about half of all IF Metall workplaces where they should be appointed (Gellerstedt, 2012; IF Metall, 2011).

Kommunal (the municipal workers, and the largest LO union) retains a strong organization in the public sector, but the union has seen a further privatization of their jobs, mainly in health and elderly care and in public transport (L-B. Johansson, 2010). Many members at first get only temporary jobs with their new private employers. This drastically reduces their SR and other union activity. However, after such transition time, Kommunal has succeeded in rebuilding the local unions and recruited new SRs. The union therefore retains their high rate of safety reps, with 30,000 SRs for some 500,000 members (Gellerstedt, 2012). But reps are rare in some of Kommunal's private-service firms. For example, farms and firms that provide personal assistants to people with handicaps are mostly very small. Their workers usually have a personal loyalty to their employers and don't want to raise problems, despite that they often face risks at work (L-B. Johansson, 2010).

The construction industry has been much affected by the market changes, especially their small firms (Frick, 2009). Despite this, the unions retain 1 SR per 22 members. Employers and unions in the industry have also recently launched a program against illegal labor, which often (though not always) is the same as exploitation of migrant workers (BBIS, 2011). This program also aims to improve the usually poor WE management in small construction workplaces, which may help the unions to recruit and train more SRs.

The low unionization and SR rate in private services may perhaps increase in some of its subindustries. Employers in private care as well as in labor-hire want to improve their reputation and to cooperate with the unions and their SRs, which may include OHSAS 18001 (and other) certifications (J. Johansson, 2010). Fastighets (organizing cleaning and house-maintenance workers within LO) has managed to increase their number of safety reps to 1 for every 28 members, despite the social dumping in cleaning (Frick, 2009). A new collective agreement in hotels and restaurants has halved their very common use of temporary labor (LO, 2011). This should make it easier for the union to organize restaurant workers (now only one in three is a member; LO-tidningen, 2012), which may enable the union to recruit more safety reps and reduce the present rate of 79 members per rep.

The LO unions' ability to recruit new safety reps is also indicated by the years served as SRs. This did not increase in 2006 from 1995, with an average of around

60 / SAFETY OR PROFIT?

7 years (Gellerstedt, 2007: 41; LO, 1997: 58). In 1980, the average time was lower, at around 4 years, but this was after many reps had been appointed during the 1970s. The lowest time as rep (i.e., the highest recruitment of new safety reps) was then in workplaces with fewer than 50 employees (Bolinder et al., 1982: 159).

Safety Reps in Growing White Collar Work

The work environment became an issue also in the white collar unions during the 1970s reforms (Keisu Lennerlöf, 1981). However, it has not become as important as in the LO unions, judging by programs and reports (e.g., SACO, 2010) and by allocated staff. Their safety rep quota also remains much lower than in the blue collar unions. TCO unions now appoint some 31,450 SRs (1 per 38 members), and SACO's growing unions 5,000–6,000 reps (1 in 125 members; Du and Jobbet, 2012). The safety rep density is around 20% higher for the working members, as union numbers mostly include several retired, unemployed, and student members. On the other hand, SRs also represent nonunionized employees, which now are some 30% of all. With a gradually increasing labor force, the ratio for safety reps is down to 1 SR in 42 employees.

A study on the safety rep role indicated that the higher the level in the workplace hierarchy, the harder it was to recruit reps, a problem especially for the SACO unions (Imander, 1999). TCO's largest union (Unionen, manufacturing and trade) now has some 6,000 reps for 420,000 active members (Du and Jobbet, 2012). In Unionen's small firms, membership is very low, safety representatives are rare (only in 11% of the workplaces; Unionen, 2008), and the WE activity is low (Unionen, 2010a). A survey of larger workplaces, where there were Unionen SRs, found more WE activity, but with too little SR participation in decisions and growing problems of stress and overtime (Unionen, 2010b).

As is the case with the other federations, TCO has more SRs in the public sector than in the private. Its unions for nurses, pharmacists, government and municipal civil servants, teachers, and police officers have 1 rep per 23–33 members. Its private sector unions instead have 1 rep per 62–91 members (Du and Jobbet, 2012). The public sectors' central agreement contributes to its many SRs. Codetermination and work environment issues are to be integrated in the dialogue between managers and employees. Local union representatives are therefore, in principle, also safety reps. However, many who act as white collar safety reps may not have been identified as such to the employers. The physicians union estimates that only 700 of their 3,000 local representatives are SRs (per 31,000 active members; Lycke, 2011).

White collar unions have focused mostly on psychosocial problems. From the late 1990s, several of them surveyed their members' working conditions and found widespread (and often growing) stress and similar risks (e.g., Unionen, 2008, 2010a, 2010b; Lärarnas Riksförbund, 2005). They have therefore become more active in supporting their safety reps with instructions (e.g., Sveriges Läkarförbund, 2005), but still with less knowledge of the SRs' situation than in the LO unions. Several SACO unions have no safety rep register, as the reps are appointed and the work environment issues are dealt with locally (Du and Jobbet, 2012; Fristedt, 2011; Jusek, 2011).

The largest SACO union, of engineers, has some 700 registered SRs (plus many unregistered) for 128,000 members. Their members are also sometimes supported by joint white collar SRs appointed by TCO's Unionen (often organizing the same workplaces), which has a better but far from full SR coverage of smaller firms. A survey of the engineers' safety reps indicates that their members too face stress and other psychosocial problems. Dealing with products and production, they may also be exposed to technical and chemical risks. The engineers once asked the LO reps for help with such risks, but with the fracturing of large corporations, many engineers work in exclusively white collar firms, usually without safety reps. The engineers have therefore started to appoint joint regional safety representatives with some other SACO unions in the private sector to support their many dispersed members (Skagerfält, 2011). This will gradually expand the professional unions' system of regional reps as a response to the need for work environment support to members in the increasing number of small firms.

Temporary white collar jobs have become more common, as direct but temporary employment and as jobs provided through agency labor hire firms that provide more skilled and expensive labor. Some 70% of the hired labor work under white collar agreements (J. Johansson, 2010). Teachers, nurses, and other public sector white collar unions (like Kommunal in LO) have also had many jobs privatized to often small firms that are hard to organize. To appoint safety reps through the union may therefore not fit the new industrial relations reality of a sometimes very low unionization in private service firms. If, for example, only 2 out of 100 employees in an IT firm are union members, not only the employer but also other employees may not find that 1 of these 2 would be the best safety rep. Union rivalries are rare in Sweden, but in less unionized companies, several competing unions may request to appoint SRs. A low unionization may thus make the WE dialogue more complicated and costly, also for employers (J. Johansson, 2010).

Activity and Time for Their Tasks

In the 1996 survey, the total activity of LO's SRs amounted to the equivalent of 10,000 full-time positions (LO, 1997: 103). By far, most of the reps took only 1–2 hours a week for their task, but there were also 4% on around half-time and another 4% (some 2,500) who were full-time safety reps. This was much more than in LO's 1980 survey, when there were 2% half-time and 2% full-time reps (Bolinder, Magnusson, Nilsson, and Rehn, 1982: 159). There is no obvious explanation for the increased safety rep activity from 1980 to 1996, but the many new reps in the expanded SR system in 1980 may not yet have fully developed their roles and interaction with managers.

Those on half- or full-time are senior SRs in large workplaces, such as hospitals or factories. They often have a high status and much influence (Frick, 1994). LO's 2006 survey didn't ask about the SR's time, but the unions estimate that this

62 / SAFETY OR PROFIT?

activity has not changed much (Gellerstedt, 2010; IF Metall, 2011; L-B. Johansson 2010). The 37,000 SRs in TCO and SACO should add at least 1,000 full-time equivalents. These in all 10,000–12,000 "full-time" SRs can be compared with the (now not much preventative) OH services' staff of 4,500 (Kindenberg, 2011) and SWEA's staff of 556, of which some 260 are labor inspectors (AV, 2011a). There is little information on how much time managers allocate to their duty to assess and control risks at work, but a study on the administrative burden of WE regulations indicates that they spend very little on this (NUTEK, 2006), as do case studies of WE management (e.g., Frick and Forsberg, 2010). The safety reps are therefore also quantitatively a crucial actor within the Swedish work environment system. However, 4 years after the 1996 LO survey, the unions found that very many SRs were pressed not to take enough time for their task (ATK, 2000). Earlier research indicated that reps very often had to informally step in when managers didn't manage risks at work (Frick, 1994). This may be why 10,000–12,000 full-time equivalents in 1996 were seen by the unions to be too few in 2000.

However, most LO reps are fairly satisfied with their situation. In the 2006 survey, 35% of them found that they had time enough for their tasks, 42% had reasonably enough time, and 20% had time only for the most important issues, while 3% did not at all have enough time (Gellerstedt, 2007). Most of them could meet and talk with their members, but 22% reported that shift work and such hampered that. There are no data on the white collar SRs' time, but they are, even more than their blue collar colleagues, responsible for their own production tasks. This means that management not only have to give TCO and SACO reps paid leave (as is their right) but also has to organize substitutes when the reps perform their WE tasks. This is not always understood and acted upon by the employers. For example, the provinces often do not organize for physician safety reps to get time off for their tasks, including training (Lycke, 2011). Likewise, overtime is frequent within Unionen (2010b), which makes it difficult for their SRs to take much time for their WE tasks.

Support and Power Balance in the Workplace

Up to the 1990s, safety reps were important voices for the scarce labor that many growing and profitable employers competed for. At least in the large manufacturing sector, it was cheaper to combine production investments with work environment improvements than to pay for the costly exit of dissatisfied workers (Frick, 1994). But since Sweden's economic crisis in the early 1990s, unemployment has varied between 6% and 13%. There is also a steady growth in casual forms of work, with lower salaries and worse conditions (Kommunal, 2011). Labor, and their safety representatives, have thus less power versus capital.

Gellerstedt (2011) analyzed the development of job qualifications from 1991 to 2009. On the one hand, the education level rose, and there was a slight shift among both workers and white collar employees toward more qualified jobs. On the other hand, the job content was reduced within each job type. More and more respondents found that the required competence was low, time for training was short, work tasks were repeated, and there was little learning on the job. Gellerstedt illustrates the dual development with some more skilled blue collar jobs but also many unskilled jobs, in which work is often tailored and controlled by computers. Most unskilled jobs are in private and public services dominated by women. To this should be added the increased workload reported in the general WE surveys and in those by teachers and other public service unions (AV, 2010; Lärarnas Riksförbund, 2005).

All in all, Swedish managers have less labor-market need to listen to safety reps in 2012 than during the reforms years. Yet the reps' influence varies between sectors and between workplaces and individual managers. The long postwar cooperation in combining work environment and productivity may have taught many managers the importance of and possibility for such combinations. These lessons are valid also in a labor market with more employer power. This may be why LO's safety rep survey from 2006 still indicates much consensus, that is, safety rep influence. The reps answered that managers more often than not cooperate with them (Gellerstedt, 2007). According to the reps themselves,

- 56% got support from their managers, but 5% found that their managers opposed them, while 36% of the SRs answered "neither-nor."
- 44% of the SRs saw both managers and employees as active in improving the WE conditions, 23% found the opposite, while some 15%–18% regarded either the managers or the employees as too passive in this.
- 29% of the SRs made formal improvement requests, as informal cooperation had not resolved the problems. This mostly (75%) resulted in an acceptable answer from the employer. If not, 47% of these SRs (3% of all) called in the labor inspection (SWEA), which mostly supported the SRs' views.
- According to the reps' replies, there was much less WE cooperation in the 30% of the workplaces without a joint WE committee, i.e., in the small ones that still had reps. Their SRs found dialogue with management sometimes to be regular (22%), mostly occasional (50%), but often nonexistent (28%).
- 67% of the SRs reported compliance with the SWEM provisions in their workplaces. Only 44% of the single SRs found this, which supports other research that SWEM is less developed in small firms. Some 63% of the SRs reported that their workplaces have action plans to improve the WE. About 66% of the SRs, or their union, had been part of drawing up this plan (which is down from the 83% participation in 1996). And 72% of the SRs found that the plans included all necessary measures. When there were major changes, 72% of the SRs (or their union) took part in the WE risk assessments.

That the shrinking LO unions can retain their local SR structure indicates the viability of the Swedish model for local WE cooperation. But the LO survey also demonstrates that this cooperation is gradually concentrated in large workplaces— where a shrinking majority of Swedes work—and is being reduced in the large majority of smaller ones. The proportion of safety reps who report having joint WE committees has grown from 47% in 1980 (Bolinder et al., 1982), over 57%

64 / SAFETY OR PROFIT?

in 1995 (LO, 1997) to 70% in the 2006 survey (Gellerstedt, 2007). This is hardly caused by more and more committees, but rather by fewer and fewer safety reps in workplaces with under 50–100 employees (50 is when committees are required, but are far from always set up). And reps without committees consistently report less influence and worse conditions. The usual difference between "good" and "bad" workplaces seems thus to have become even wider. This gap is also indicated by more frequent union reports of safety rep harassments, especially in small firms that are much exposed to unfair competition from migrant firms and/or illegal work, such as in construction (Gellerstedt, 2010). The growth of small firms without safety reps has, as mentioned, made the white collar unions increase their system of regional safety reps, however these remain far fewer than in the LO unions. For 2010, LO reported the equivalent of 232 full-time RSRs, TCO 40 RSRs, and SACO 5 full-time regional reps (AV, 2011b).

Not only management but also union and member support is important in promoting the safety reps' influence on the work environment. There may be a problem of internal union competition between safety reps and negotiators. Still, 85% of the reps in LO's 2006 survey had good support from both members and the union, though 41% found local cross-union cooperation wanting, with other LO unions or with those from TCO or SACO (Gellerstedt, 2007).

Contested Issues and Social Construction of Risks and Solutions

Conflicting Views on the Definition and Regulation of Psychosocial Risks

Safety rep influence is not only determined by managers' willingness to listen to them, but also by how the SRs define and act on the risks and solutions that they want managers to listen to. This is the power of deciding the dialogue's agenda, emphasized by Lukes (1974). How risks at work are socially constructed is a result of both conscious decisions and the development of work. The issues of work environment debate have changed with the shift from blue to white collar work. Technical risks are far from eliminated. With the fracturing into smaller firms and more foreign firms with posted workers, the prevention of technical risks may also deteriorate (Frick, 2009; and as mentioned by the engineers union above). Nevertheless, organizational factors causing psychosocial and musculoskeletal health effects have come to dominate as causes of work related ill-health (AV, 2010).

These risks are more diffuse than the technical ones. SWEA have issued provisions for some problems—harassment, threats and violence, and working alone—but these do not clearly state what is permitted or not. Even the attempt to specify psychosocial requirements is contested. SWEA started to draft general psychosocial provisions in 2002 but withdrew the draft after employer critique for interference with their freedom to manage production (Bruhn and Frick, 2011). Unions therefore find that the agreement on stress between the EU's social partners has not been implemented in Sweden and that safety reps lack adequate instruments to promote a better psychosocial work environment (Frick, 2010). Labor inspectors and safety reps (or anyone) who want to reduce organizational risks instead have to make do with the SWEM provision's general duty to assess and, if necessary, act against all risks (Frick, 2011b). This has led to many employee surveys, especially in larger workplaces. However, when these, for example, reveal widespread sleeping problems from stress, there are no provisions to support safety reps' requests to reduce the workload (Frick and Forsberg, 2010).

Training and Other Enlightenment to Support the Work Environment Dialogue

Psychosocial risks are little regulated, complex, and widespread. It is therefore no surprise that 77% of LO's safety reps wanted training on this in the 2006 survey (followed by ergonomics, 70%; Gellerstedt, 2007). This is a sharp increase from the 1980 and 1996 surveys, when 34% and 39%, respectively, of the reps wanted such training, and technical risk training was more urgent (Bolinder, Magnusson, Nilosson, and Renn, 1982: 164; LO, 1997: 102). However, in ATK's evaluation (2000), psychosocial training was emphasized also by the LO unions, and it has long been the focus of white collar unions' SR training (Keisu Lennerlöf, 1981).

Training is an essential right for safety representatives. Their role in local "voluntary" improvement and compliance rests on reps having good arguments in their organized dialogue with managers. Managers also have to be informed by the employers, as mandated in, for example, the SWEM provisions. The abolition of state subsidies for SR training and for other enlightenment measures in 2007 (after earlier reductions) may therefore hamper the local WE cooperation. All information and training of safety reps is now to be paid by their employers. The unions find that this has led to a serious reduction in the SR training, even worse than described by them in the ATK report (2000). Local unions and employers have a joint responsibility to train reps. This is to be paid for by the employers, but unions have often not only found them unwilling to do so but also reluctant to give the reps paid time off to be trained (LO-tidningen, 2008a). An inventory within the LO unions in 2005 revealed that some 60% (44,000) of the safety reps lacked work environment training. Despite the withdrawn subsidy, the unions managed to train some 20,000 of these reps during 2005–2007 (LO-tidningen, 2008b). And Prevent— the social partners' joint training and information office—sold some 10,000–15,000 of its basic WE course handbook plus 15,000–25,000 other materials per year during 2000–2010, materials which often are (re)used to train new groups of safety reps and others (Quist, 2011).

Still, many safety reps lack training. LO's 2006 survey showed that 37% of the responding SRs had had no training, and 24% had at most one day during the last year. And 44% of the SRs believed that their supervisors/managers had some WE training, but 43% didn't know. Only 27% of the SRs had been trained jointly with their supervisors/managers. When there were local WE agreements or joint

66 / SAFETY OR PROFIT?

WE committees, only 18% dealt with training for safety reps and others (Gellerstedt, 2007). After the LO survey in 2006, the state subsidy was withdrawn for the further, usually central, training of reps. The participants in LO's central WE training dropped from some 1,200 yearly during 2000–2006 to around 300 in 2007–2009. However, LO and the private employers (Svenskt Näringsliv) have agreed to use 10 million SEK per year during 2010–2014 to support further SR training, which has brought this up to the old levels for LO's SRs (Sundberg, 2011). But SACO had to close its central SR training in 2007 (Fristedt, 2011), and TCO provides considerably fewer central WE courses (Sjöholm, 2010).

The state's enlightenment strategy, of a shared and informed view on risks as means to improve work, has also been reduced in other ways. The abolition of the National Institute for Working Life in 2007 closed by far the largest provider of such R&D, information, and training. The closure of SWEA's safety rep register (due to the authority's huge budget cuts) reduced the overview of the SRs' situation. The (growing) white collar unions especially often lack such registers of their own. They have to make do with assessments of how many reps they have in various types of workplaces (Fristedt, 2011; Jusek, 2011; Lycke, 2011; Skagerfält, 2011). The OH services have, as mentioned, also lost much of their role as prevention advisors after their government subsidy was abolished in 1993.

More Individual and Professional Perspectives on Work and its Risks

Unions support the role of safety rep through training, instructions, advice, and meetings. Still, reps spend by far most of their time in their ordinary jobs. Senior safety reps on half- or full-time may identify much with this role, but other reps' perceptions of work and its risks are mainly shaped by their work groups' communities of practice (Gherardi and Nicolini, 2000a, 2000b; Wenger, 1998). The understanding of work and the work environment learned within these groups affects the safety reps' views and attitudes and also the support that reps can get from their peers. Reps cannot act too much out of tune with their group or their peers may regard them less as one of us (representing us) and more as one of them (those above who decide; Sjöström, 2006).

The shift toward more professional and individualized jobs is a major change in the basis for how work is constructed in the communities of practice. With Sweden's strong manufacturing base, a typical employee during the work environment reforms of the 1970s was a male factory worker in a group of peers under direct instruction of a supervisor. Now, the typical employee is almost equally often a woman and mostly a white collar employee in either private or public services, who is managed by objectives and with customer responsibility. The work environment (and the safety reps) has become important also for the white collar unions, but these tend to see work and risks within the broader setting of their members' professional situation and career development (Jusek, 2011; Lärarnas Riksförbund, 2005). The LO unions still have many taylorized blue-collar jobs, but now they are mainly controlled through computerized monitoring systems. And blue collar unions also promote members' professional and individual development to retain their jobs and get higher salaries (Gellerstedt, 2011).

Many skilled white collar employees (e.g., physicians; Lycke, 2011) also experience the unlimited production pressure described by Lysgaard (1961). Professionals who retain more control over their work situation may also be exposed to unhealthy stress by internalizing too-high production objectives (Levi, 2001). With internalized work demands more than external supervision, and with individualized jobs, white collar employees, but also many workers, have less similarity in their tasks; factors that Lysgaard saw as necessary for the social construction of a worker's collective to jointly set defensive work norms against the unlimited pressure.

There are no studies to clearly indicate what such changes may entail for the task of safety representatives. However, some possible effects may be

- White collar SRs are not only less common than in blue collar jobs, but professional roles and internalized production goals may also make it harder for them to get their colleagues' support for their work environment task, as was reported already by Keisu Lennerlöf (1981).
- The cohesion and solidarity within Lysgaard's (1961) worker collective may, at least partly, be replaced by professional identity and solidarity, e.g., as teachers or physicians but also as electricians and other skilled workers. This should depend much on if work environment risks are an inevitable part of the professional discourse and on how this can be supported by union training for safety reps and other members.
- The motivation and loyalty of more skilled, or even academically trained, employees with more personal production responsibilities are important for the employer. Inasmuch as work risks are part of the social construction within the various jobs' communities of practice, this should give the safety reps an internal bargaining power that is less dependent on the external labor market. On the other hand, skilled and responsible jobs may also lead to the mentioned internalization of employers' production objectives. Work environment critique and requests, by safety reps or any employee, may then be difficult, as it is a critique of one's own work.

New HRM and Risk-Management May Reduce Employee Representation

The position and the social construction of the task as safety rep have thus become more complex. The reps remain in need of union training and other support, as described by Walters and Nichols (2007). However, this union construction of the safety rep role—as distinct from the job/professional role—faces different issues than in the blue collar goods production with authoritarian management. Modern management methods also present a challenge to uphold the role as critical safety rep against the employee production role. Such management methods, often inspired by Japanese quality-control principles, may reduce collective cohesion and solidarity as employees and increase identification with the company and its production. Even in highly unionized Sweden, the spread of lean-production management may reduce the employees' collective identity and union loyalty (S. Johansson, 2011). Swedish workplaces also have more international owners, often used to more adversarial and less cooperative industrial relations (Levinsson, 2004; Wikman, 2010).

Risks and prevention are also politically constructed. Besides the much reduced (but far from abolished) enlightenment strategy, there is a political search for causes and cures for societal problems. The debate on health at work has, since around 2002, gradually focused more on the individual and less on working conditions as causes and on individual measures and behavior change as cures. This may affect how safety reps perceive risks and prevention.

- The growing (but still very imperfect) implementation of the mandatory SWEM may erode the old activist role of safety reps. If managers start to be more proactive against (at least some) risks at work, employees may go to them instead and turn less to their safety reps. In the similar Danish setting, Lund (2002) and Dyreborg (2011) analyzed how more active managers may reduce the safety reps' influence. And in Australia, occupational health and safety management has even been used as a means for union busting (Frick, 2004).
- The politics of sickness absenteeism at first focused on the work environment, especially stress, as a major cause. With little evidence, the debate turned to the alleged misuse of sickness insurance as the main problem (Johnson, 2010). This individual focus has been the basis for many projects to reduce sickness absenteeism, but with less interest in risk prevention. In the public sector, employers use sickness absence as the main work environment indicator, despite evidence that this is a poor measure of risk exposure (Frick and Forsberg, 2010).
- The focus on the individual more than on risks at work is also evident in the return of behavioral safety as injury explanation. This is now getting popular among managers, also in (some) advanced manufacturing companies (Rasmussen, 2010).

CONCLUSION: RESILIENCE OF SAFETY REPRESENTATIVES IN AN INCREASINGLY DIVISIVE LABOR MARKET

How Swedish safety reps have been affected by the changed economy is complex and ambivalent. It is therefore not advisable to deal in generalities. However, it is possible to draw some main conclusions from the developments taking place at the three levels discussed above.

The Formal Position, Rights, and Coverage of Safety Reps

The proportion of safety reps is down from around 1 in 31 employees in 1980 to 1 in 42, though their absolute number is still around 100,000. Behind this figure is a change in the structure of the SR system:

- The rate of safety reps per member in the three union federations is roughly the same as in 1980, or 1 SR in 20 LO members, 1 in 38 in TCO, and 1 safety rep for around 125 SACO members. However, the labor market has grown, and its composition has shifted. LO, with its blue collar jobs, has shrunk from 63% in 1975 to 45% of the union members in 2010, while the white collar jobs in TCO's unions grew slightly, and the academic and professional jobs organized in the SACO unions a great deal.
- All federations have fewer members and safety reps in the growing number of small firms, especially in private services. LO's 65,000 reps are now even more concentrated in larger workplaces and employers. Within the growing white collar federations, this development has made more SACO unions start to cover small workplaces with regional safety reps (although they are still far less active in this than the LO unions).
- Safety reps have recently gotten the power to represent agency labor, but other proposals to counter the diffusion of work environment responsibility in the increasing supply chain economy have not been enacted.

Workplace Power Balance and Safety Rep Influence

Safety reps' labor-market power is clearly reduced, but the effect of this on their dialogue with managers depends on several factors, on which there often is a lack of data. After less activity during the 1990s, unions have become more active in supporting their safety reps in the face of a decreased membership and a more fractured workplace structure. The safety representatives' answers in the LO survey of 2006 describe a mostly acceptable and sometimes quite good WE dialogue. However, 20%–30% of them still reported various problems. Some 36% of them found the managerial support to be weak, and 5% saw themselves resisted in their task by management. Although the SRs mostly get good support from their unions and members, they also often lack a cross-union cooperation.

There is also a positive bias in the LO survey. The large majority of the answering reps are in larger workplaces, with several SRs. Far too few single reps are appointed in smaller workplaces, and the single SRs report many more problems in performing their tasks. Unions also report increased harassment of single safety reps. The gap seems thus to have increased between better (nearly always larger) and worse workplaces in WE cooperation. Most Swedes still work for large employers, but more and more are employed in small firms. The changes in job structure and content (Gellerstedt, 2011) also indicate a growing difference between better and worse jobs, and jobs with less control and content reduce the influence safety reps have on managers.

The Contested Social Construction of Risks

Unions try to support their reps with training, not the least on psychosocial risks. There is less attention to possible problems in the safety rep role due to the increase of a more active risk management and through sometimes very low unionization, that is, when only a few employees are union members. The prevention dialogue as a basis for the safety reps' influence may also be challenged by new management methods of lean production, HRM management, and also by more focus on the individuals' responsibility to safeguard their own health at work.

In all, this has resulted in some deterioration of WE cooperation and of the numbers of SRs and their influence, but less so than could be expected from the weakened position of labor. There are also some mitigating tendencies. Blue collar jobs in private services and construction have been disproportionately exposed to these changes, especially in small firms. The unions have still been fairly successful in upholding their SR system in these, but even more in other industries. The overall changes have therefore mainly increased the dual labor market with wider differences between core workers and those on the periphery. And who is a core worker to the employer can differ much between industries and firms. However, enlightenment as a strategy to voluntarily reduce risks and improve health at work has been reduced for all. Both blue and white collar unions see a shortage of training for their SRs, as was already evident from the LO survey of 2006. Less training and information results not only in lack of local knowledge of risks and how to prevent them, but it may also erode the social norms that support WE prevention. The combination of a wider gap between "good" and "bad" workplaces and the erosion of knowledge and norms is a long-term threat to the Swedish cooperative work environment system, based on active safety representatives.

http://dx.doi.org/10.2190/SOPC4

CHAPTER 4

Old Lessons for New Governance: Safety or Profit and the New Conventional Wisdom

Eric Tucker

New governance theory has a large following in academia and is influential in numerous spheres of regulatory policy (see Davies, 2011, for a critical discussion). Yet in the area of occupational health and safety, new governance is hardly new at all. Indeed, it is fair to say that in many ways what are now labeled new governance concepts were first articulated and applied in the 1972 Robens Report, *Safety and Health at Work*. This included its critique of command and control legislation and its emphasis on the development of better self-regulation. In Robens' words (1972: para. 41):

> *The most fundamental conclusion to which our investigations have led is this. There are severe practical limits on the extent to which progressively better standards of safety and health at work can be brought about through negative regulation by external agencies. We need a more effectively self-regulating system* (emphasis in original).

Thus, it is particularly fitting that we return to Theo Nichols and Peter Armstrong's early critique of the Robens Report for some old lessons for new governance in occupational health and safety (OHS) regulation (Nichols and Armstrong, 1973). While much of their monograph criticized the Robens report (1972) for blaming apathy as the underlying source of workplace injuries (which justified its call for more self-regulation), Nichols and Armstrong's response was that a proper understanding of risk creation had to take as its starting point the pressure for production generated within capitalist relations of production. From their political economy perspective, the central regulatory problem was how to counteract the pressure to prioritize production over safety, and their solution required shifting power over production to workers on the shop floor.

72 / SAFETY OR PROFIT?

Clearly, the terms of the debate over OHS regulation have not remained static in the nearly 40 years that have passed since their intervention; nor have the economic, political, and social conditions of production. Thus, the goal of this chapter is to follow the growth and development of new governance thinking about OHS, but at the same time, and in the spirit of Nichols and Armstrong, to critically examine the underlying assumptions new governance theorists make about the world and the implications for their prescriptions if they are wrong, and to consider alternative reforms that focus on building public regulatory capacity.

The chapter proceeds in four parts. In the first, I construct some ideal types of OHS regimes based on three variables; state protection, worker participation, and employer management systems. These are used as heuristics in subsequent discussion. The second part briefly discusses the roots of new governance in the Robens (1972) report (referred to as "old" new governance) and briefly reviews Ontario's experience with it, to illuminate its dynamics and its vulnerability to regress toward neo-liberal self-regulation/paternalism in the absence of effective worker OHS activism. In part three, I focus on recent work by two North American new governance theorists, Orly Lobel and Cynthia Estlund, who consciously wish to avoid a collapse of new governance approaches into neo-liberal self-regulation/paternalism. I argue that despite their aspirations, the new governance prescriptions they embrace are unlikely to be institutionalized with the protective conditions they advocate and that their emphasis on self-regulation valorizes a movement toward the destination they wish to avoid. Finally, I ask whether degradation toward neo-liberal self-regulation/paternalism is inevitable and, if not, whether a progressive new governance theory that aims to strengthen a regime of public regulation under the unfavorable conditions that prevail today provides a better alternative.

CONSTRUCTING IDEAL TYPES

For the purpose of locating new governance theories of OHS regulation within a range of possible configurations, it will be useful to construct some ideal types of regimes. In the past, I attempted to map out regimes of OHS regulation using two axes, state protection and worker participation (Tucker, 2007: 145–170). The focus of that study was on worker citizenship in the OHS regimes, not with OHS regulation more generally.

That approach left out an important dimension of current regulatory practice, promotion of employer OHS management, which must be brought in if we are going to investigate new governance's emphasis on self-regulation. The term *OHS management* as used here does not refer specifically to the presence or absence of an occupational health and safety management system defined as a "systematic managerial process to detect, abate and prevent workplace hazards" (Frick, Jensen, Quinlan, and Wilthagen, 2000). Rather, it refers to *state policies* that support the development of employer competence and commitment to manage OHS (other than by command and control regulation or strengthening worker participation). This might include education and promotional activities, support for the formation of sectoral safety associations, and the use of economic incentives in the workers compensation system, including experience and merit rating.

For the purposes of constructing ideal types, I have adopted a binary weak/strong assessment for each element, although obviously this is a gross oversimplification that ignores the enormous variation within actual regimes based on industry, region, and such. Table 4.1 presents the eight logically possible combinations.

I am not particularly concerned about defending the labels I have attached to these ideal types, but perhaps it is worth briefly explaining my thinking on them. The regimes can be divided into two clusters based on whether there is weak or strong state protection. Beginning with the former, in a laissez-faire or neo-liberal regime, the state does not actively support any of the three dimensions of OHS regulation. In this regime, it is unlikely that employer provision of high quality work environments will spontaneously emerge, notwithstanding the business-case arguments that are often made for it (for a literature review of the business case, see Davis, 2004: 69–72; Dorman, 2005: 351; Hart, 2010: 585; Johnson, 2005). The historical evidence for laissez-faire regimes that dominated much of the 19th century supports this view; workers were routinely exposed to highly dangerous conditions in mines, factories, and railways, and suffered high rates of injury, disease, and death (Rogers, 2009; Tucker, 1990). A similar conclusion can be reached about OHS management in a

Table 4.1 Ideal Types of OHS Regimes

	Laissez-Faire/ Neo-liberal Regime	Paternalist Promotional Regime	Worker Participation Regime	Collective Laissez-Faire Regime
State Protection (Command and Control Regulation)	Weak	Weak	Weak	Weak
Worker Participation	Weak	Weak	Strong	Strong
Employer OHS Management	Weak	Strong	Weak	Strong
	Command and Control Regime	Command and Promote Regime	Workers' Democracy Regime	Social Democracy Regime
State Protection (Command and Control Regulation)	Strong	Strong	Strong	Strong
Worker Participation	Weak	Weak	Strong	Strong
Employer OHS Management	Weak	Strong	Weak	Strong

74 / SAFETY OR PROFIT?

neo-liberal world when union representation and state regulation are declining and production is increasingly organized through diffuse contractual networks rather than direct management of employees (Nichols and Tucker, 2000: 285; Quinlan and Mayhew, 2000: 175; Woolfson and Beck, 2003: 241).

For this reason, most states have chosen to intervene on at least one of the three dimensions of regulation. The regime that deviates least from the ideology and practice of laissez-faire or neo-liberalism operates by promoting only employer OHS management, typically through some combination of educational activities, support for the formation of employer safety associations, and the use of economic incentives in a workers' compensation system. I have labeled this kind of regime "paternalist promotional," insofar as it does not provide workers with any rights and operates primarily by encouraging management to behave responsibly. Arguably, this approach prevailed in the regulation of farmworker safety in Ontario before the occupational Health and Safety Act was extended to agriculture in 2006. Farmworkers neither enjoyed an entitlement to minimum OHS standards nor a right to participate in OHS management, but they were covered by the workers compensation system, which supported a farm safety association, and farm employers were experience rated and so had economic incentives to reduce claims costs. Not surprisingly, this OHS regime performed poorly (Tucker, 2006b: 256).[1]

A third regime in this cluster is one built primarily on strong worker participation rights that enable workers to exert a significant level of influence on and, perhaps, control over decision making that affects health and safety conditions at work. Strong worker participation regimes must be distinguished from weaker involvement schemes that are likely to be largely cosmetic and aim to secure worker compliance with management objectives (Taylor, 2001: 122–138). It is possible that strong worker participation will emerge in the absence of active state support when there is widespread and effective worker self-organization, but it is more common for participation rights to become generalized and entrenched through legislation requiring safety representatives and joint health and safety committees (JHSC).[2] It is difficult, however, to think of any state which has constructed an OHS regime primarily around strong worker participation.

A more common route for states that wish to limit direct state regulation but hope to improve OHS outcomes is one I have labelled a collective laissez-faire regime. The term is borrowed from Kahn-Freund's characterization of the English

---

[1] On the exclusion of farmworkers generally from labor protection in the United States, and on the limited role of the federal Occupational Safety and Health Administration, see Schell, 2002: 139, esp. 149–150. Arguably, a regime of this sort operated in Ontario after the enactment of a workers compensation law in 1914 and successfully promoted OHS improvements (Silvestre, 2010: 527).

[2] For example, in Ontario miners negotiated JHSCs before they became mandatory at a time when direct state regulation of health hazards in the mines was quite weak. However, worker OHS representation became widespread only after the enactment of mandatory legislation (Tucker, 2007: 151).

regime of labour law, which he viewed as based on state support for the establishment of collective bargaining processes that allowed the parties maximum leeway to conclude and enforce their own agreements.[3] In the context of OHS regulation, the focus is on state support for bipartite regulation through the promotion of worker participation rights (including forms of collective representation) and employer safety associations. While it is difficult to identify a pure form of this model, as we shall see in the next section, regulatory bipartism has been attractive to some Robens-style OHS regimes (Tucker, 2007).

The second cluster of OHS regimes is based on strong state protection. A pure command and control model relies exclusively on direct state regulation. Although early health and safety laws fit this model in principle, the reality was that OHS standards were almost always qualified by considerations of what was practical, and enforcement practices were based on persuasion rather than prosecution (Tucker, 1990). Arguably, then, many of these regimes operated more in a paternalist promotional mode. However, to the extent that they combined strong direct state regulation with active support for employer management, they would fall into the model I have labeled a "command and promote" regime. Finally, I have labeled a regime that combines strong command and control regulation with support for worker participation a workers democracy regime, and a regime that combines a strong version of all three a social democratic regime. Scandinavian models tended toward the last, as they supported strong employer OHS organization in conjunction with worker participation rights to promote bipartite regulation with a strong state presence. (Tucker, 1992: 95).[4]

It is important to reemphasize that this is a table of ideal types, not actual historical or current examples of OHS regulation, which are always going to be more complex. Nevertheless, I think it provides some useful heuristics for thinking about and comparing actual OHS regimes and in particular, for understanding and ultimately assessing both "old" and "new" new governance approaches.

"OLD" NEW GOVERNANCE

As we noted, new governance theory in the realm of OHS regulation first became prominent in the Robens Report with its emphasis on the development of more effective self-regulation. The role of the state was to create the conditions for this to occur (Robens, 1972, para. 41). However, as I have argued elsewhere, Robens-inspired regimes of mandated partial self-regulation were implemented in very different ways, depending on the degree of self-regulation permitted and on the substance of what was mandated, particularly in regard to requirements for worker participation. The political and economic context in which the regulation operated

---

[3] For a recent discussion of Kahn-Freund's approach, see Dukes (2009: 220).

[4] An illuminating discussion of the difference between collective laissez-faire and a more thoroughly social democratic model of regulation, as represented by the work of Hugo Sinzheimer, is provided by Dukes (2008: 341).

also significantly shaped the capacity and willingness of workers to exercise their participatory rights and the strength of the state enforcement effort. Finally, the flexibility inherent in these regimes meant that they could be reconfigured without the necessity of new legislation by administrative action and changes in worker and union leverage at the point of production. The result was that the regime was always a political project in the making, subject to the competing objectives of workers and employers and therefore likely to experience recurring conflict over its design and implementation, at least as long as workers were actively engaged in pushing for stronger protection or participation (Tucker, 1992: 95, 2003: 295, 2007: 145).

In Ontario, the Ham Report (1976) called for legislation to promote internal responsibility systems (IRS) that delineated the responsibilities of employers, supervisors, and workers and that granted workers rights to know about workplace hazards, to participate in OHS management through joint health and safety committees (JHSCs), or, in smaller workplaces, through safety representatives, and to refuse unsafe work. The IRS was to be the primary site of regulation, with the external responsibility system (ERS) of enforceable standards, inspections, and prosecutions to provide a backup in instances in which the IRS failed. In short, Ham's recommendations aimed to institutionalize a regime that most closely resembled collective laissez-faire, albeit one in which worker participation rights were rather limited.

The legislation that followed, however, did not simply enact the Ham Commission (1976) recommendations. Rather than being an apolitical measure implementing expert recommendations, the entire exercise, from the appointment of the Commission to the enactment of legislation, was shaped by conflict between a strong worker OHS movement that was pressuring the government to address serious OHS hazards that it was publicizing on an ongoing basis and an employer lobby that was resistant to encroachments on managerial prerogatives. The resulting OHS legislation was shaped by these pressures, but also aimed to contain them (Storey and Tucker, 2006: 157–186). It promoted self-regulation through the IRS with a secondary role for direct state protection. This approach did not calm the political waters, as worker OHS activists often found management resistant to addressing their health and safety concerns and the state reluctant to enforce the law. In response, workers called for legislation to increase their power within the IRS and to strengthen the ERS—a regime more akin to the worker democracy model (Fidler, 1985: 315; Smith, 2000; Storey and Tucker, 2006). More pressure on the government led to legislative revisions in 1990 that better institutionalized worker participation by mandating more procedural requirements for JHSCs, but did not expand the powers of committees or increase the strength of worker rights. The law also expanded regulatory bipartisanism by providing for worker/ management oversight of health and safety training in the province. Finally, maximum penalties for violations were substantially increased. Overall, there was some movement toward a social democratic model, but the extent of the change was modest.

Ironically, the election of a labor-friendly New Democratic Party government in 1990, shortly after the enactment of the OHS reforms, was accompanied by less, not more, enforcement. Moreover, the militant worker OHS movement that had been so effective over the previous 15 years or so began to weaken, perhaps in part because trade union officials did not want to embarrass "their" government and in part because the economic recession of the early 1990s not only made other workplace issues more pressing but also increased the level of job fear. In many ways, it began to look as if Ontario was going to go down the path followed in the UK under New Labour (Tombs and Whyte, 2010a: 50–61) but, as we shall see in the final section of the chapter, it did not.

The lessons from Ontario's experience of "old" new governance accord well with Nichols and Armstrong's (1973) political/economy analysis. First, worker OHS activism was absolutely essential to the initiation of change. Second, that activism significantly shaped the reform legislation; in the absence of political pressure from workers, the Ontario regime would have been more paternalistic, in that the mandated IRS would provide for weaker worker participation and state enforcement. For example, in the interim legislation, passed in 1976, JHSCs were only required when ordered by the ministry. Worker OHS activism pushed the model incrementally closer toward the workers' democracy or social democratic model. Third, continuing worker OHS activism was necessary to prevent the regime from degrading into a neo-liberal/paternalist regime, in which health and safety would be constructed according to management perspectives on what was reasonable given the cost constraints under which management operated (profit over safety). Instead, activist workers were able to turn JHSCs into arenas in which independent worker voices exerted some influence on management decisions (Hall, Forrest, Sears, and Carlan, 2006: 408). Fourth, continuing worker OHS activism was responsible for pushing the government to strengthen worker participation rights, even if only in the direction of better institutionalizing worker voice in the IRS and the ERS but without giving workers more power. Finally, the election of a so-called labor-friendly government did not guarantee that labor-friendly policies and practices would be implemented. As a result, continued worker mobilization and pressure were essential in preventing regulatory backsliding. Finally, as worker OHS activism declined, the regime slipped back toward paternalism.

"NEW" NEW GOVERNANCE

The Robens Report's (1972) emphasis on the limited capacity of the state and the need to promote more responsible self-regulation lies at the heart of new governance theory, which has since been elaborated and given additional theoretical justifications. When addressing the limits of traditional command and control regulation, new governance theorists emphasize the growing complexity of the external environment and the increasing speed of technological and organizational innovation. This, they argue, overloads the capacity of states to gather and process the information necessary to develop effective regulations. They also emphasize that

78 / SAFETY OR PROFIT?

processes of globalization are hollowing out the ability of national states to effectively regulate globalized activities and that supranational regulatory institutions are too underdeveloped to fill the void. Moreover, they posit that command and control strategies are self-defeating. The imposition of substantive standards creates rigidities both in law and in the social fields that regulation is attempting to control. Existing regulation is likely to become quickly outdated and irrelevant, producing both the phenomenon of overregulation of old hazards and underregulation of new and emerging ones. Finally, the targets of regulation increasingly resist what regulation there is and oppose new regulation, thus further limiting the state's capacity to respond to changes in the work environment, rendering it even less effective. In short, they say, we have a vicious circle. Not only does command and control regulation impair the capacity and motivation of private actors to solve problems on their own and reduce economic efficiency, but it also fails to achieve its normative goals. Workers and employers are both made worse off.[5]

Much of this narrative might also fit within a neo-liberal critique of traditional regulation, but new governance theorists insist they are not abandoning the normative goal of improved OHS outcomes to the naked pursuit of efficiency within an increasingly competitive global economy. Rather, they claim that new governance theory provides an approach to regulation that steers a third way between command-and-control regulation, on the one hand, and neo-liberalism on the other. A variety of names have been appended to this approach, including reflexive regulation, responsive regulation, and regulated self-regulation, among others (see, for example, Ayres and Braithwaite, 1992; Teubner, 1983: 239). Although there are variations in emphasis, the focus of new governance. is on steering corporate governance or management systems in socially desirable directions—other than by simply commanding them to behave in prescribed ways. It is posited that law can facilitate more cooperative processes that are flexible, responsive, and participatory, and that align the firm's interests with substantive regulatory goals that are not fundamentally altered by new governance methods. Tools include the use of incentives, increasing participation of stakeholders, requiring information sharing, and formal auditing, to name a few. A focus on networks and the identification of nodal points at which influence can be exerted, often by other private actors, is central to this approach (Braithwaite, 2008: for a critical discussion, see Davies, 2011).

People who have been involved in OHS regulation for the past several decades might wonder, with good reason, whether new governance is worthy of the appellation "new." Not only are the roots of these ideas found in the Robens Report (1972), but OHS regulation in many countries, including the UK and, as we have seen, Canada, has developed more or less along the regulatory track that new governance theorists are now elaborating, including mandatory disclosure of hazard information, mandated JHSCs, and a reduced reliance on state standard setting and

---

[5]The literature is vast. I have tended to focus on North American contributors, particularly Lobel and Estlund, whose work will be discussed in more detail below.

enforcement. For this reason, it is not surprising that new governance theorists have been attracted to current practices of OHS regulation as outstanding, if imperfect, examples of new governance theory in action.

In this regard, it is worth spending some time looking at Lobel's influential 2005 article, "Interlocking Regulatory and Industrial Relations: The Governance of Workplace Safety" (Lobel, 2005: 1071) since it provides insight into both the strengths and weaknesses of this kind of approach. Much of the article is devoted to documenting the failure of U.S. OHS regulation, pointing out that, as a system of command and control, it is incredibly ineffective: inspection resources are grossly inadequate, penalties for violating the law are paltry, and prosecutions are rare. Moreover, the new world of work is increasing the difficulty of regulation because of accelerating innovation, vertical disintegration of firms and the corresponding growth of complex supply chains, and the challenge of addressing injuries that arise not from trauma but from job stresses and strains. What then is to be done? The overarching objective is to make OHS a shared interest that can be integrated into the core ends of economic enterprises. Regulatory agencies can achieve this by identifying the conditions under which effective self-regulation works, promoting their development, and recognizing when those conditions are absent (Lobel, 2005: 1104).

Taking the case of the U.S. Occupational Safety and Health Administration (OSHA), Lobel identifies a number of voluntary compliance programs developed in the 1980s and 1990s that signalled a shift away from command and control to a new governance model that fostered public/private partnership, encouraged industry cooperation, and allowed flexibility in policy implementation (Lobel, 2005: 1111). However, she is also keenly aware that such a shift can also provide a cover for neo-liberal deregulation by allowing firms to escape even a weak threat of sanction for failure to achieve politically established regulatory standards (Krawiec, 2003: 487). Therefore, Lobel (2005) argues, institutional arrangements must be put into place to avoid this result, and she identifies two principles, exit and voice, to guide this exercise. The exit principle requires the creation of a dual-track system of enforcement and sanctions so that firms are faced with a clear choice: if they fail to responsibly self-regulate, they will be made subject to an effective, coercive command and control regime of enforcement. The threat of the big stick has to be credible. The voice principle requires effective worker participation in the firm's occupational health and safety management system. This fits with new governance's emphasis on mobilizing nongovernmental actors as a means for pressuring private corporations to act responsibly, even if this is not in the corporation's short-term economic interest.

On the face of it, Lobel's (2005) prescriptions resemble a social democratic model of OHS regulation insofar as she insists on the importance of a strong state to deal with laggards, worker participation in the design of regulated self-regulation schemes, and state support for employer OHS management. Indeed, in earlier work, she argued quite strongly for the view that new governance should be committed to the achievement of social democratic values (Lobel, 2004: 387). For that article,

80 / SAFETY OR PROFIT?

Lobel identified three overarching projects intertwined in new governance theory: economic efficiency, political legitimacy, and social democracy. Lobel noted that choices must be made and balances struck between these projects, but explicitly adopted a social democratic perspective, insisting that substantive commitments to the achievement of public ends must be maintained (Lobel, 2004: 387). But how difficult will that be?

The answer will depend to a great degree on the view taken of regulatory dilemmas in OHS and their resolution (Tucker, 2010: 99). In the political economy tradition, exemplified by Nichols and Armstrong (1973), OHS regulation is seen as a realm of recurring regulatory dilemmas that stem from the relentless requirement within capitalism to produce for profit. To do this, capitalists must constantly pursue technological and organizational innovation. This drive does not always lead to the creation of hazardous working conditions, but historically, it often has and presently it often does. But the fundamental point, as Nichols and Armstrong argued, is that there is systematic pressure within capitalist economies to privilege profit seeking over other objectives, including OHS, whenever those other objectives impose a barrier to the circulation and expansion of capital. The development of regulation *for the benefit of working people*[6] involves the imposition of limits on the freedom of owners and managers of capital to engage in profit seeking at the expense of safety. As such, regulatory dilemmas will be a recurring phenomenon and must be considered from a class perspective, since at its core, OHS regulation involves a conflict between those who want to impose limits on capital and capital's interest in maintaining maximum freedom to manage and control its activities in a self-interested way.

Erik Olin Wright's model of the conditions for class compromise suggests that positive class collaboration is possible within certain limits when workers have achieved a certain level of power (Wright, 2000: 957). Apart from situations in which improved safety and profit maximization coincide, the fundamental condition is that working class organization must be sufficiently strong so that the costs to employers of cooperating with labor are lower than the costs of conflict. This can happen because worker organization, if it is sufficiently strong, can help employers overcome their collective action problem by, for example, imposing standardized conditions across an industry and reducing competitive pressures. Applied to OHS regulation, the argument would be that if workers, operating in conjunction with the state, can insure that *all* employers operate at the same high standard, employers will be more inclined to co-regulate the work environment since they can be assured that their competitors will not gain an edge by producing less safely but more cheaply.

---

[6]It is worth noting here that in *Regulatory Capitalism*, Braithwaite (2008) emphasizes the ongoing role of regulation and therefore rejects the view that a neo-liberal order has emerged. What he ignores, however, is that the nature of regulatory capitalism has changed. Instead of regulation that restrains capitalism, we now have regulation that facilitates the growth of economic and social inequality. To make the persistence of regulation the central story while ignoring the shift in class power that neo-liberal policies have facilitated seems peculiar at best.

Swedish models of OHS regulation in the 1970s and 1980s arguably went some way toward meeting this condition, but as Swedish capital has become more globalized in recent decades, workplace organization at the national level is less able to reduce competitive pressures (Frick, 2009; Tucker, 1992: 95).

New governance theorists are not insensitive to the existence of conflicts of interest and unequal power relations but, like "old" new governance theory, they generally seek to minimize their salience and avoid the need for trade-offs. The assumption of common interests in OHS deeply informed the Robens Report. "Indeed, there is a greater natural affinity of interest between 'the two sides' in relation to safety and health problems than in most other matters. There is no legitimate scope for 'bargaining' on safety and health issues, but much scope for constructive discussion, joint inspection, and participation in working out solutions" (Robens, 1972: para. 66). It is notable that Robens put the terms "the two sides" and "bargaining" in quotations seemingly to emphasize his doubt that there really were two sides that had something to bargain over (Robens, 1972: para. 13, 28, 41). The report recommended worker participation, not to act as a check on management's penchant to stint on OHS, but rather to encourage workers to accept "their full share of responsibility" and to monitor conditions for the purpose of providing information that employers would act upon (Robens, 1972: para. 59). This view also informed the Ham Report (1976), which asserted that health and safety was not a suitable issue for collective bargaining.

Recent new governance theorists have continued in this tradition of emphasizing common interests and marginalizing the significance of class conflict over OHS. Davies (2011) emphasized the ways in which governance and network theorists reject the continuing salience of class and class-based conflict. For example, Lobel states, "Nonetheless, rich bases of ethnographies and comparative studies indicate that intense conflict between labor and capital around issues of occupational health and safety may be anomalous." She then quotes Robens for this proposition (Lobel, 2005: 1128).[7] Earlier in her article, Lobel also addressed this issue in developing an argument that firms operated pursuant to a mixed set of motivations. However, even within that context, Lobel emphasized the business case for safety, taking into account the reduction of accident costs, including improved worker morale, reduced absenteeism, and consumer preference for goods and services that are produced safely (Lobel, 2005: 1102).

This last point links to an argument often found in "new" new governance theory about the significance of reputational risk as a driver of firm behavior. Here, the argument is that a poor health and safety performance will negatively affect the firm's bottom line either because consumers will avoid its products or because other firms will not wish to do business with unsafe firms. Safety pays, at least most of the time, reducing the need to confront trade-offs between economic efficiency

---

[7]Notably, the additional sources in her footnote on this point (fn. 274) do not for the most part provide empirical support for the common interest proposition.

(profit) and social democratic values (worker safety). The assumption of common interests between workers and employers built on the business case for safety has a long history in OHS regulation going back to the factory acts and has often provided the basis for a subtle renorming of its objects toward the elimination of those risks that are excessive—from a business perspective (Tucker, 1990: 167–173). Notwithstanding the persistence of business-case arguments, often found on government websites, the empirical evidence supporting the business case has been subject to much criticism and its limits identified.[8]

In addition to discounting the extent to which OHS conflicts with the profit motive, new governance theorists also tend to minimize the strength of the profit motive when conflicts do arise. In particular, there appears to be a high level of faith in the likelihood that virtue will triumph over self-interest. For example, in a recent discussion of restorative justice in the OHS context, Braithwaite describes restorative justice as a process that is

> about sitting in a circle discussing who has been hurt and then the victim being able to describe in their own words how they are coping with the hurt and what they are looking for to repair that harm and prevent it from happening again. It is about the virtue of active responsibility as opposed to passive responsibility of holding someone responsible for what they have done in the past (2008).

If no one takes responsibility, the circle is widened until a "softer target"— presumably the virtuous actor—is hit (Braithwaite, 2008: 76–77).

Lobel also picks up on this strain of the new governance theory when she argues that "a cooperative governance framework can create empathy and mutual trust among diverse people" (Lobel, 2005: 1104). Left here, we might think of this as an argument that virtue is immanent in business actors and just needs a little nudging to be activated and to become the guiding principle for business decision making. However, at least one empirical study has cast doubt on whether restorative justice responses influence behavior (Nielson and Parker, 2009: 376). This result should not be surprising if one takes into account the ways in which unequal power relations in hierarchical organizations marked by real differences in motivations and goals of upper-level managers and lower-level workers undermine communication and may lead to the construction of hegemonic norms that shift responsibility from management to individual workers (Silbey, 2009: 361–363; Zoller, 2003: 118). Moreover, as Harry Glasbeek notes,

> Given the corporation's recent historical role in the jettisoning of such job and income security as had been won . . . why should anyone believe that there will be a corporate drive to give back some of these gains? There is a limit to

---

[8]For an example of government promotion of the business case strategy, see http://www.hse.gov.uk/betterbusiness/large/index.htm

the extent that managers can indulge their personal sense of altruism and/or worker friendliness and still be true to their real task (Glasbeek, 2008: 203).

And that real task, as corporate law scholars will tell you, is maximization of shareholder value, with all it entails (Ireland, 2005: 49).

Because Lobel minimizes the salience of conflict and regulatory dilemmas, the OHS policy prescriptions she favors are problematic. Lobel endorses OSHA's efforts to promote cooperative compliance through a variety of 2-track programs that exempt qualifying firms from regular inspections, although she recognizes that in the absence of a strong commitment to command and control regulation for firms that do not qualify for preferred treatment, the model is likely to "become a guise for deregulation" (Lobel, 2005: 1114). As well, she is critical of the failure of OSHA's compliance programs to systematically include workers' voices and believes this too should be required. Her solution, then, includes elements of state protection and worker participation, but she provides little concrete guidance about how these should be structured, particularly in the context of pervasive and growing labor-market inequalities. Indeed, her treatment of inequality is particularly disappointing, since it is entirely abstracted from capitalist relations of production (Lobel, 2005: 1141–1144). Against this background, Lobel emphasizes cooperative programs and employer OHS management, with state enforcement and worker participation as contributors to the goal of better employer self-regulation. Yet she fails to confront the implications of prioritizing this track in a world in which these necessary supports are unlikely to materialize in the absence of concerted efforts to rebuild, legitimize, and strengthen public enforcement and worker participation.

Here we might turn to Cynthia Estlund's (2010) recent book, *Regoverning the Workplace*. Estlund comes to new governance theory and regulated self-regulation as much out of despair as hope. She sees the death of old forms of workplace regulation as irreversible and more self-regulation as inevitable. Therefore, the question for her is whether it is possible to steer the development of self-regulation toward new forms of workplace governance in which workers have a real voice. Her book might be read as an internal dialogue in which she tries, with only partial success, to convince herself that this is possible.

Her considerable reservations about new governance prescriptions derive from her better understanding of and greater focus on the realities of unequal power relations in the workplace than Lobel's. For example, although Estlund treats OHS as an area in which there are sometimes common interests and argues that common interests are more likely to exist in OHS than, for example, minimum wage or hours of work laws, she also recognizes that there will be occasions when "hazards . . . are integral to the production process and . . . serve the employer's bottom line" and that sometimes "health and safety improvements come with a significant price tag" (Estlund, 2010: 178).

Because of a greater sensitivity to salience of power imbalances in the workplace, Estlund recognizes the vulnerability of employee committees to cooptation and intimidation, given the fear of retaliation and job loss, even for unionized employees

(Estlund, 2010: 176).[9] Moreover, she also recognizes that internal committees are unlikely to effectively address hazards when they conflict with profit making. So while safety committees can assist in some ways, for example, by aggregating and articulating employee knowledge about hazardous conditions when lack of communication is part of the problem, they need to operate in conjunction with some outside entity "that can supply power, independence, and protection against reprisals" (Estlund, 2010: 179).

Unfortunately, Estlund's proposals to meet that need are not very promising. Corporate codes of conduct with private independent monitoring and/or employee whistle-blower protection are proffered as part of the solution. The other piece is a 2-track enforcement regime built on a beefed-up enforcement track for firms that fail to self-regulate with the promise of a second track of less frequent inspection and reduced penalties for firms that have codes of conduct with independent monitors and protected worker participation. Expanded rights of action for workplace injuries, retaliation for whistle-blowing, and violations of health and safety laws firm this up.

Much has been written about the efficacy of corporate codes of conduct backed up by independent monitors in the context of global supply chains. Apart from questions about the independence of monitors and their effectiveness, there is a more fundamental concern with the logic of the model. In nearly every case, the motivation is the protection of reputational capital, and this is most heavily concentrated in highly branded firms that face a real threat of consumer pressure. While NGOs might be able to mobilize consumer opinion around the Nikes of this world, a broader dependence on voluntary and privately monitored codes seems unlikely to reach vast areas of the economy in which consumer branding does not play such a large role.[10] Legislative measures imposing legal liabilities backed by meaningful inspection will often be required to achieve effective supply chain regulation (Walters and James, 2011: 998).

The alternative of depending on workers to blow the whistle on their employers is also problematic. Most health and safety acts currently protect workers against retaliation for exercising their statutory rights, including the right to refuse unsafe work, as well as to make complaints, but in the absence of a strong union, most workers are unlikely to be assertive protagonists. This is amply illustrated by Neil Gunningham's work on the Australian mining industry, which draws a very pessimistic view about the effectiveness of worker participation in a cold industrial relations climate (Gunningham, 2008; Quinlan and Johnstone, 2009; Walters and Nichols, 2007). This returns Estlund (2010) to the conundrum of requiring the presence of the very conditions she stipulates as absent to make her alterative

---

[9]For a recent discussion of growing insecurity among workers with permanent, full-time work and its negative effects on their health, see Lewchuk, Clarke, and de Wolff, 2011.

[10] For skeptical views, see Arthurs (2009) and Locke (2013).

effective. As she recognizes, in their absence, neo-liberal or paternalist regimes of regulation are far more probable outcomes than social democratic ones.

To the extent that the response to this weakness is the presence of a strong enforcement track that would induce firms to sign up to avoid facing its teeth, it presupposes the existence of one of the conditions whose absence has been identified as a driver of the turn to regulated self-regulation. If regulators lack the capacity both to identify which firms genuinely are on the right track and to detect OHS violations and appropriately sanction firms on the wrong track, regulated self-regulation is quite likely to degrade into neo-liberal self-regulation/paternalism.

Finally, Estlund's faith in the power of private litigation to create incentives for employers to comply with minimum standards laws is, perhaps, uniquely American. She is critical of the lack of avenues for private litigation for workplace injuries and OHS violations and calls for "activating the prodigious regulatory forces inside firms" by "arming . . . workers themselves . . . with their own regulatory arsenal" in the form of private actions (Estlund, 2010: 233–234). Although it is true that there has been an increase in employment litigation in the United States, it is hard to imagine that most workers, let alone vulnerable workers, are likely to benefit from private rights of action or that they will be an effective substitute for adequate public compensation and enforcement regimes.

So, while there are differences among "new" new governance theorists, the dominant tendency is to dissolve safety/profit conflicts by assuming common interests prevail and to minimize the extent to which profit-seeking behavior will take precedence over safety considerations when conflicts do arise by assuming that virtue will tend to trump self-interest. While more social democratic new governance theorists are quite cognizant that under conditions of unequal power relations new governance techniques face severe challenges and could be used to further disempower the weak, their call for safeguards still leaves in place a core agenda that favors more self-regulation and reliance on nonstate actors. Moreover, they have already found that the safeguards they insist upon—state enforcement and worker voice—are irretrievably in decline. As a result, their shift in emphasis, from public to private systems of regulation, may actually increase the likelihood that enforcement deficits will be exacerbated rather than redressed.[11] In that vein, Davies (2011) argues that the promotion of new governance theory can be fruitfully understood as a dimension of a neo-liberal hegemonic project that aims to secure consent through participation and the promotion of common interests, while the practice of new governance fails to deliver a new cooperative social order because it is undermined by the neo-liberal material conditions in which new governance practices are enacted.

---

[11] For another critique of new governance theory applied to enforcement deficits, see Davidov, 2010; Jain et al., 2010; Marshall, 2010: 555 (indicating that a reflexive regulatory scheme has failed to improve the condition of Australian outworkers).

ALTERNATIVE PATHS

Clearly, one path of new governance practice is toward more self-governance, with the real danger that it will produce neo-liberal or at best paternalist regimes. As Tombs and Whyte (2010a: 46) have demonstrated, this is the path that has been followed in the UK. From Estlund's (2010) perspective, the self-regulation train has left the station, and so there is no alternative but to work within that framework and try to steer it toward more worker protection. If that is true, then indeed our options are limited, and the prospects for regulatory renewal are dismal. The question I want to pose here is whether there are alternative paths, and if so, whether a progressive new governance theory has anything to offer. My answer to both these questions is a tentative yes; that a regime that maintains a strong public enforcement focus is still possible and that a theory that focuses on ways to mobilize civil society forces in aid of public enforcement can indeed make a contribution to the development of effective OHS regulation in the context of the new world of work.

First, in regard to the possibility of regulatory alternatives that retain a strong public enforcement component, let us return to the case of Ontario. As we saw earlier, at the beginning of the 1990s, it was looking as if the regime of mandated partial self-regulation was going to degrade as state enforcement was ebbing and worker OHS activism was subsiding. Subsequent events, however, tell a more complicated story, one that points to the possibility that the trajectory of public enforcement is not necessarily downward sloping, even under unfavorable political and economic conditions.

In 1995, an ideologically right-wing government was elected, and it attacked collective bargaining and employment standards laws. But it did not go after OHS. Indeed, the enforcement effort actually began to intensify, and this upward trend has continued since the election of a Liberal government in 2003, although it has dipped slightly in the past couple of years (Tucker, 2003: 409–412). Evidence of this can be seen in the number of inspections, orders issues, stop-work orders, and prosecutions (see Charts 4.1–4.3). Of course, some of this increase may reflect the desire of government to pump up numbers. It is easy enough to manufacture the appearance of enforcement intensification by having more superficial inspections and issuing more trivial orders, but this explanation is belied by a concomitant increase in the use of stronger enforcement powers, notably stop-work orders and convictions.

The news is not all good. The number of convictions, particularly after 2005, is inflated by the expansion of a ticketing system that allows inspectors to lay on-the-spot charges against workers, supervisors, and employers for a variety of OHS violations. In practice, workers and supervisors each receive about 37% of summonses, while employers receive 25%. As Garry Grey notes (2009: 326), the targeting of workers and frontline supervisors blurs the definition of who is an OHS offender and diffuses responsibility. It also has resulted in a reduction in the average fine per conviction (see Chart 4.4).

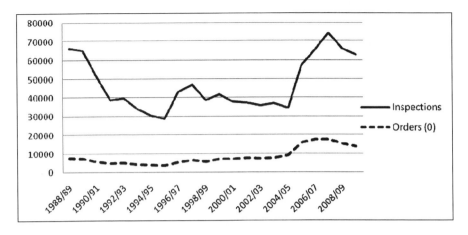

Chart 4.1 Ontario Ministry of Labour, OHS Inspections and Orders Issued 1988/89–2009/10.
Source: Ontario Ministry of Labour, Enforcement Statistics, various years.

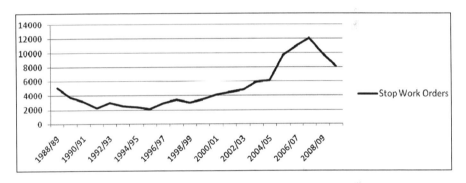

Chart 4.2 Ontario Ministry of Labour, OHS Stop-Work Orders Issued 1988/89–2009/10.
Source: Ontario Ministry of Labour, Enforcement Statistics, various years.

In order to assess trends in more serious prosecutions under Part III of the *Provincial Offences Act*, disaggregated data was obtained from the Legal Services Branch for three years, 2007–2008 to 2009–2010. The data show that while there has been an increase in Part III prosecutions over these years, from 369 to 445, the average fine per conviction for these more serious offences declined from $35,303 to $28,839.[12]

---

[12] Special data run, Legal Services Branch, April 2011 (in possession of author).

88 / SAFETY OR PROFIT?

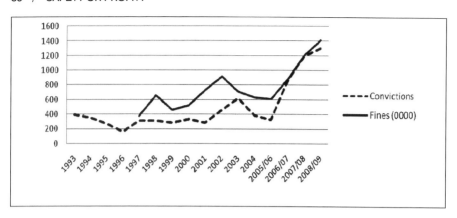

Chart 4.3 Ontario Ministry of Labour, OHS Convictions and
Annual Fines Issued 1993–2008/9.
Source: Ontario Ministry of Labour, Enforcement Statistics, various years.

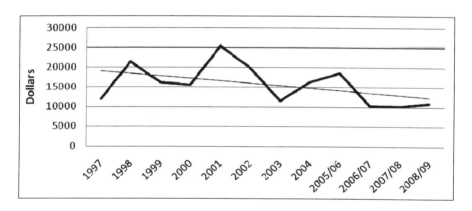

Chart 4.4 Ontario Ministry of Labour, OHS Convictions,
Average Fine per Conviction, 1997–2008/9.
Source: Ontario Ministry of Labour, Enforcement Statistics, various years;
calculations by author.

It is also noteworthy that there has been little use of the criminal sanctions notwithstanding the enactment of Bill C-45, the so-called Westray Bill, which came into force in 2004 and was supposed to facilitate prosecution of OHS crimes.[13] Across Canada, fewer than 10 criminal charges have been laid. Three criminal cases have been brought in Ontario. The first criminal prosecution (*R.v Fantini* [2005] O.J. 2361 (ONCJ)) arose out of a ditch collapse that killed a worker in Newmarket,

---

[13] For a critical assessment of the legislation that has arguably been borne out by its application, see Bittle, 2013 and Bittle and Snider, 2006: 470.

Ontario. The charge was laid in 2004 and was resolved by a plea deal in which the criminal charge was dropped in exchange for a guilty plea to violations under the OHS statute (Edwards, 2005). The second criminal prosecution in Ontario was launched against Millennium Crane, the company's owner, and the crane operator at the time. In this case too, the charges were dropped after an engineering report failed to support the prosecution's case (Canadian Employment Law Today, 2011).[14] The third prosecution arose out of a scaffolding collapse that killed four immigrant workers on Christmas Eve 2009. The accused included the Metron Construction Co. and three of its officials (Keith, 2011). The company pleaded guilty to criminal negligence as was fined $200,000 plus a 15% victim surcharge. The crown, which had sought a 1 million dollar fine, appealed but at the time of writing the outcome is unknown. Criminal charges against the president of the corporation, Joel Swartz, were dropped and he pleaded guilty to four charges under the OHS Act and was fined $90,000 plus a 25% victim surcharge (*R. v. Metron Construction Corporation,* 2012 ONCJ 506; *R. v. Swartz,* 2012 ONCJ 505). Criminal charges against the project manager are still pending (Edwards 2013).

In Quebec, two criminal cases have resulted in convictions. In Transpavé Inc., involving a workplace fatality, the accused pleaded guilty and was fined $110,000 (*R. v. Transpavé Inc.,* 2008, QCCQ 1598). The first conviction after a trial was obtained against Pasquale Scrocca, a landscape contractor. In that case, an employee died when the brakes on a backhoe Mr. Scrocca was driving failed, pinning the employee against a wall. Scrocca received a conditional sentence of imprisonment for 2 years less a day to be served in the community (*R. v. Scrocca,* 2010 QCCQ 8218). One case, *R. v. Gagné* (2010 QCCQ 12364) ended with an acquittal. In that case, charges were laid following a collision between a train and a maintenance vehicle, which resulted in one death and three injuries. The two accused individuals were employees of Québec-Cartier: Steve Lemieux was the train operator, and Simon Gagné was a foreman. Justice Dionne found that the mistakes made by the employees arose from a corporate culture of tolerance and deficient training, not wanton and reckless disregard for the lives and safety of workers on the part of the accused. In principle, this finding could have resulted in a conviction of the corporation, but it had not been charged.[15] There is at least one case still pending, against Mark Hritchuk, the service manager at a car dealership where an employee died after catching fire due to a broken fuel pump (Dubowski, 2010).

Finally, in British Columbia, the United Steel Workers launched a private prosecution in 2010, arising out of the death of a Weyerhauser employee in British Columbia in 2004. The company was previously assessed a penalty of nearly $300,000 by the BC compensation board. A court has ruled that the union

---

[14] http://www.employmentlawtoday.com/ArticleView.aspx?|=|&articleid=2450

[15] Norman Keith and Anna Abbott, "Acquittal in Quebec Bill C-45 Charges" (Online at http://www.gowlings.com/KnowledgeCentre/enewsletters/ohslaw/htmfiles/ohslaw20110427.en.html)

presented enough evidence for the case to go forward, but the Crown subsequently intervened to terminate the prosecution (OHS Insider, 2011a, 2011b).

The targeting of workers and low-level supervisors in both OHS and *Criminal Code* prosecutions, and the failure to aggressively use the criminal law should not obscure the positive side of the enforcement data, which reflect an increase in inspectors generally, and of proactive inspections in particular, that target high-risk firms and through planned safety blitzes on high-risk hazards.[16] However, workers can never be complacent about the gains they make in improved enforcement. Since the beginning of the recession in 2008, which has hit Ontario particularly hard, there has been a dip in enforcement activity, and there is a risk that, as the government moves to trim its budget, enforcement, resources will be lost. There has already been a shift in employment standards enforcement, toward a self-reliance model that places more responsibility of workers to pursue their claims with their employers as part of the government's open-for-business agenda.[17]

The report of the Expert Advisory Panel on Occupational Health and Safety (Dean Report, 2010) did not make strong recommendations on enforcement, seeking instead to accommodate the concerns expressed to it by employer and labor stakeholders. Thus, it called for "a consistent approach of tough enforcement for serious and wilful contraventions, as well as compliance assistance where guidance and support for employers help achieve compliance." The report called for a review of the ticketing system, with an eye toward increasing set fines and for the addition of administrative monetary penalties as an enforcement tool.[18] It remains to be seen how the government will respond to these recommendations.

Data on the strength of worker participation is harder to come by, but what little there is suggests that it has weakened over the past decade or so. One study comparing IRS systems in 1990 and 2001 found that worker influence seemed to be in decline. Managers viewed worker participation as less important and perceived that management bargaining strength over OHS had increased. Worker representatives saw management as less cooperative and perceived that workers were more likely to be hassled by co-workers for raising health and safety issues or filing a workers compensation claim (Geldart, Shannon, and Lohfeld, 2005: 227–236). Another possible measure, albeit one that is subject to alternate interpretations, is the number of work refusals reported to the Ministry of Labour.[19]

---

[16] For a description on Ontario's OHS enforcement strategy, see http://www.labour.gov.on.ca/english/hs/sawo/index.php (accessed April 13, 2011). For a study that supports the efficacy of a blitz strategy in the context of workplace discrimination see Hirsh, 2009.

[17] Open for Business Act, S.O. 2010, c. 16, Sch. 9, s. 1. For a critique, see Workers Action Centre and Parkdale Community Legal Services, Submission to the Standing Committee on Finance and Economic Affairs regarding Schedule 9, Bill 68, An Act to promote Ontario as open for business by amending or repealing certain Acts (July 26, 2010). Online at http://www.workersactioncentre.org/!docs/sb_Bill68_eng.pdf (accessed October 14, 2011).

[18] For a more detailed discussion of the Dean Report, see Lewchuk, this volume Chapter 8.

[19] For an insightful analysis of work refusal behavior, see Gray, 2002.

To the extent that they tell us something about worker OHS activism, the trend over most of this decade has been downward (see Chart 4.5). The recession of 2008 undoubtedly has exacerbated worker fear of job loss and dampened their willingness to be militant around OHS issues.

The decline in private sector union density in Canada over the past 3 decades, from nearly 30% in 1981 to just over 16% in 2009, would be consistent with decreased worker OHS activism of all kinds, given the significant link researchers have found between union representation and the effectiveness of worker participation (for example, Lewchuk, Leslie, Robb, and Walters, 1996; Nichols and Walters, 2009). The drop in union density also translates into reduced bargaining power and militancy, as is reflected in the sharp drop in strike incidence since 1976 (see Chart 4.6).

The Dean Report recognized that more needed to be done to support worker participation in the IRS and made a number of recommendations, including a measure to permit worker chairs of JHSCs to submit written recommendations to the employer when the JHSC is deadlocked, mandatory training of health and safety representatives in small (6–19) workplaces, and improved protection from reprisals (Dean Report, 2010: 28–31, 49–51). These recommendations were acted upon by the government.

Assessing the strength of employer OHS management is even more difficult. One study found that, on several measures, OHS management improved between 1990 and 2001. Senior managers were more directly involved in safety, safety was more likely to be included in managerial job descriptions, and more safety training was provided to workers (Geldart et al., 2005).[20] It is also possible that increased experience rating has created an economic incentive to better manage safety, although the evidence is decidedly mixed, with studies showing that the

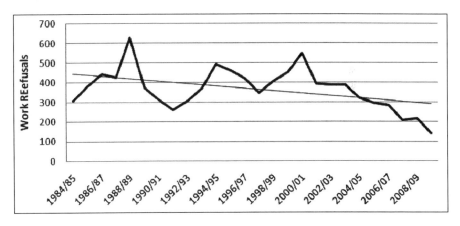

Chart 4.5 Ontario Ministry of Labour, Work Refusals Reported 1984/85–2009/10.
Source: Ontario Ministry of Labour, Enforcement Statistics, various years.

---

[20] A related study found that firms with better OHS management had lower lost-time injury rates (Geldart, Smith, Shannon, and Lohfeld, 2010: 562).

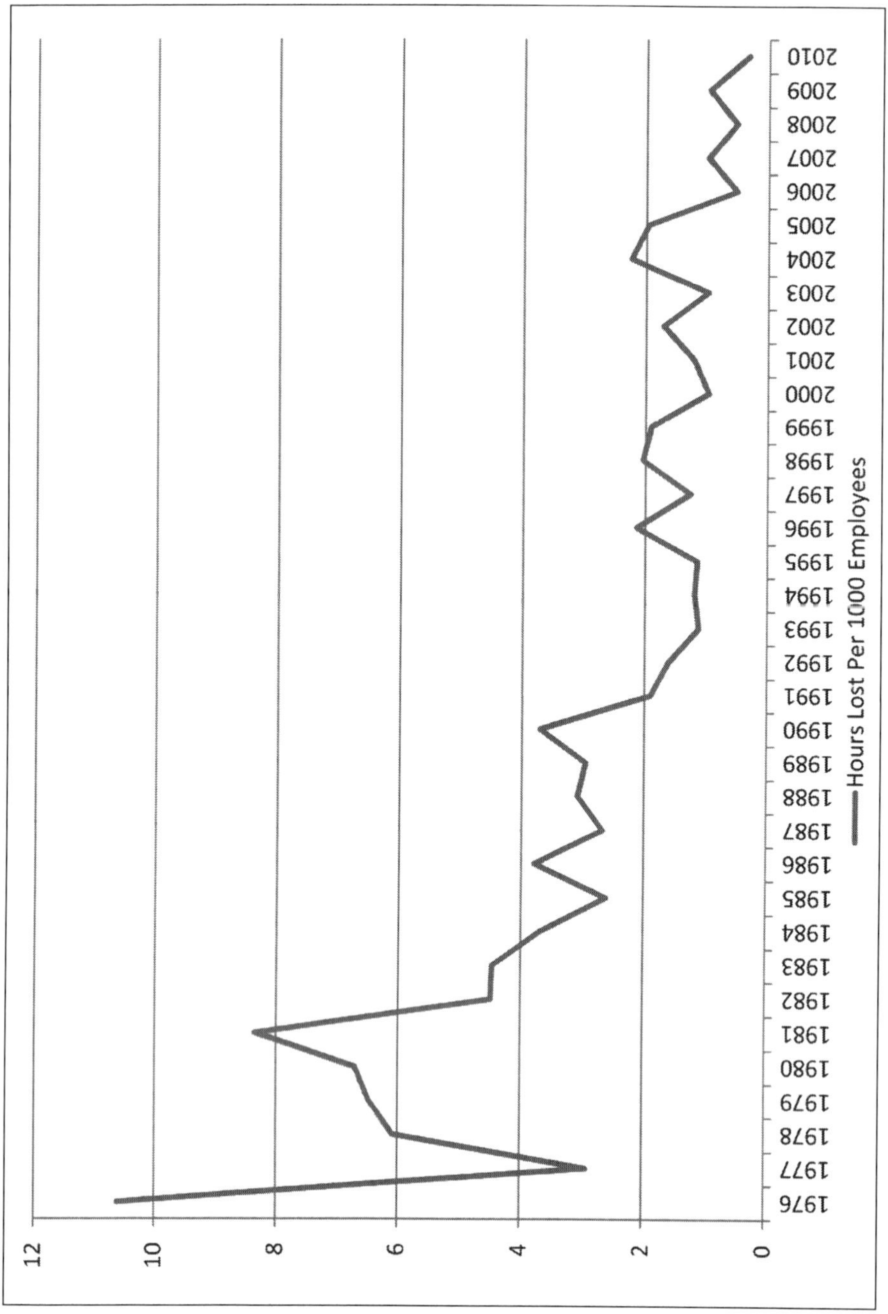

92 / SAFETY OR PROFIT?

Chart 4.6 Annual Average Hours Lost to Labour Disputes, Canada, 1976–2010 (per employed worker).
Source: HRSDC, http://www4.hrsdc.gc.ca/.3ndic.1t4r@-eng.jsp?iid=14

most frequent response of employers is to control claims cost, often through aggressive claims management (Campolieti, Hyatt, and Thomason, 2006: 118; Tompa, Trevithick, and McLeod, 2007: 85). On the other hand, the growth of small business and decentralized production is likely to have had a negative impact on the overall capacity of employers to manage OHS (Eakin, 1992: 689; Quinlan, 1999).

The Dean Report also made some recommendations to enhance employer capacity, competence, and commitment to manage OHS risks. These include mandatory OHS training for supervisors responsible for frontline workers, an accreditation system for employers that successfully implement OHS management systems, appropriate financial incentives for firms that qualify suppliers based on their OHS performance, and more compliance assistance from inspectors. It also made recommendations aimed at improving compliance in the small business sector. Importantly, none of the proposals are offered as alternatives to strong enforcement and worker participation. For example, there is no suggestion that accredited employers should be exempt from routine inspections, and the report is clear that more compliance assistance "should not interfere with the inspector's duty to enforce the law." Moreover, the report also recognizes that financial incentives should not be based primarily on claims cost and frequency (Dean Report, 2010: 53).[21]

The relevance of Ontario's recent OHS regime lies in the fact that it provides evidence that neo-liberal self-regulation/paternalism is not the inevitable destination of Robens-style regimes. Indeed, it indicates that after a regime starts down the neo-liberal path, as was the case in Ontario (ironically, under the auspices of a labor-friendly government), its direction can be changed, even with a conservative government in power. It is beyond the scope of this intervention to endeavor to provide an explanation for this development,[22] other than to suggest that perhaps in some way the assignment of prevention activities to the Workplace Safety and Insurance Board, where a discourse of partnership and "safety pays" predominates, created more space for the MOL to define its mandate as that of setting and enforcing standards. If that is true, then the principal recommendation of the Dean Report (2010) to move prevention activities back into the Ministry of Labour, a recommendation that the government has already implemented, may pose the risk of a reorientation away from enforcement. Moreover, the recent dip in enforcement activity is a timely reminder of the difficulty of maintaining a strong program of proactive enforcement in the face of a recession, which has resulted in significant job loss, particularly in the manufacturing sector. (For recent data, see HRSDC, 2011.)

This brings us back to the final question I want to address, which is whether a progressive new governance theory has anything to offer. Here, I turn briefly to recent work by Gordon and Fine, who start from the same premise of folks like Lobel and Estlund—that existing forms of command and control regulation are deficient

---

[21] On March 3, 2011, the government introduced Bill 160 to implement some of the Dean Report's recommendations.

[22] Elsewhere, I tried to explain divergences between selected Canadian jurisdictions based on trade union density, employer size, etc. (Tucker, 2003).

94 / SAFETY OR PROFIT?

and that moving forward will require the mobilization of civil society forces. However, unlike Lobel and Estlund, they reject the turn toward self-regulation as inevitable or desirable. Rather, their project aims to engage workers as monitors in *public* enforcement (Fine and Gordon, 2010: 552). Their particular concern is the enforcement of labor standards, particularly hours of work and minimum wages, and this dictates a focus on vulnerable workers in the nonunion sector—the paradigm case of the new world of work. Given a world in which state enforcement is chronically underresourced, unions have limited reach, and the incidence and distribution of complaints is unlikely to reflect the pattern of actual violations, (Weil and Pyles, 2005: 59) their recommendations aim to develop a system of third-party worker representatives who can extend the reach of public enforcement into sectors of the labor market in which workers are at high risk of violations but also less well protected. These third-parties could be unions with roots in a particular industry, but could also include worker centers or other worker organizations. For worker organizations to have a meaningful impact on enforcement, Fine and Gordon stipulate that collaborations with government must be formalized, sustained and vigorous, and adequately resourced. They suggest a number of models, ranging from deputizing persons associated with worker organizations to designating worker organizations as sites where workers can anonymously register complaints to having worker organizations provide enforcers with intelligence who would assist them in identifying high-risk employers or industries.

In some ways, their proposal resembles the Swedish regional safety representative system, except that the focus here is much less on providing assistance to the IRS than on strengthening the ERS. It is also a model that can be used by unionized workers in their own workplaces. For example, the Canadian Auto Workers recently reported on the efforts of a local that represents federally regulated workers in the airline industry to build relationships with federal OHS inspectors so that they can more easily be called upon to intervene when the union is unable to secure satisfactory responses from their employers (Bennie, Bachl, and Dewey, 2010: 3).

The approach advocated by Fine and Gordon faces many obstacles and may not be realized except on an exceptional basis. Nevertheless, the priority given to building pubic enforcement and enhancing worker participation—that is, toward building a worker democratic regime—arguably has more potential in the long run than a "new" new governance alternative that relegates these elements to a secondary and supporting role in a regime that gives primacy to employer self-regulation.

CONCLUSION

For the most part, "new" new governance theories are rooted in the same consensus theory that informed "old" new governance theory as articulated by the Robens Report (1972) and that was so effectively criticized by Nichols and Armstrong (1973) nearly 40 years ago. Lobel (2005) and Estlund (2010) are sensitive to the salience of power imbalances and conflicts of interest and therefore call for robust worker participation and public enforcement as adjuncts to a regime of regulated self-regulation. However, for both, worker participation and public enforcement are truly secondary. Lobel justifies a focus on self-regulation by minimizing the extent of conflict over OHS, while Estlund claims there is no alternative and therefore efforts must focus on steering self-regulation toward more effective worker protection. I argue their policy prescriptions are more likely to lead toward neo-liberal self-regulation/paternalism than toward worker or social democracy. The example of enforcement in Ontario suggests that the trend toward more self-regulation is not inevitable and that Gordon and Fine's strategy of enhancing *public* regulation and enforcement through partnerships with worker organizations is feasible and promising. Of course, there are significant limits to what can be accomplished in an increasingly neo-liberal capitalist social formation, which exacerbates the structural pressure to put profit over safety, but there is ample evidence that employers who are inspected and sanctioned for OHS violations become more actively engaged in workplace prevention (for a recent review, see Purse and Dorrian, 2011). Indeed, the authors of one study conclude that

> to confine the role of prosecution to a measure of last resort, then, is not only without empirical foundation, but also likely to send the "wrong message" to employers about how the problem of serious OHS offences is understood by regulatory authorities and in civil society more broadly (Schofield, Reeve, and McCallum, 2009: 275).

In short, instead of new governance approaches that depend on common interest or the mobilization of market forces by civil society groups, we are better off developing strategies that concentrate on strengthening worker voice and public regulation.[23] At the very least, such an approach avoids valorizing a set of assumptions and practices that, as Nichols and Armstrong (1973) demonstrated, historically have served workers poorly.

---

[23] David Weil's work on enforcement is particularly helpful. For example, see Weil, 2010.

http://dx.doi.org/10.2190/SOPC5

CHAPTER 5

Safety, Profits, and the New Politics of Regulation

Steve Tombs and David Whyte

INTRODUCTION: SAFETY OR PROFIT

The "common interest" principle—the key idea in which Western liberal traditions of social regulation are embedded—can be traced back to early forms of social regulation in the 19th century (Carson, 1985; Tucker, 1990). It is the same principle that formally underpins the UK Health and Safety at Work Act. As the Robens Report, which laid the foundations of the Act, asserted, "There is greater natural identity of interest between 'the two sides' in relation to health and safety problems than in most other matters. There is no legitimate scope for 'bargaining' on safety and health issues" (Robens, 1972: para. 66).

Nichols and Armstrong's 1973 pamphlet *Safety or Profit* emphasized a crucial connection between the highly questionable assumption that most "accidents" were the result of worker apathy and the assumption that both workers and managers have a common interest in overcoming this apathy through effective health and safety practice. This connection emerged from their analysis of the Robens Report (1972). And it is from this very same connection that the *business case* for safety is drawn: the idea that self-regulation can be viable, since it is ultimately in management's interests to organize effective safety practices; worker apathy is best dealt with by good management; and the principle of maximizing worker safety can coexist harmoniously with the principle of maximizing profits.

Yet Nichols and Armstrong (1973) carefully set out—through a contextual analysis of five accidents ("Four Fingers and a Sore Head") in one workshop employing some 70 people—how these incidents were caused not by apathy, or even by poor management, but by the simple fact of workers being "under pressure to keep the job going" (p. 6); that is, "the concrete reality of getting the job out" (p. 30). *Safety or Profit* documented how the fundamental reality of working life is

97

at odds with Robens' portrayal of the workplace. The empirical analysis of the organization of safety in the pamphlet shows clearly how safety and profits stood in opposition to each other, just as the interests of management and workers are fundamentally antagonistic. For Nichols and Armstrong, it was safety *or* profits, not safety *and* profits.

We wish to return here to the significance of these two small words—the significance of linking "safety" with "profits" via an *and* or an *or*. In this chapter, we examine—in the close to 40 years since *Safety or Profits* was published—how the common sense of safety *and* profits has been fundamentally re-created and bolstered. This chapter does so by examining the ways in which the business case for worker safety was restated and reinstitutionalized during the recent period of Labour Governments in the UK, and setting out the extent to which it still rests upon a Robens' "common interest" assumption (1972).

THE CHANGING WORLD OF WORKER SAFETY

Before we proceed to reexamine the common sense of safety and profits in the context of the more recent politics of regulation, it is necessary to acknowledge some fundamental changes in the British economy, and in particular, in the labor market, which both have and appear to have occurred since Nichols and Armstrong's 1973 text.

Nichols and Armstrong cite the Robens Report's 1972 conclusion that "Every year, something like 1,000 people are killed at their work in this country. Every year about half a million people suffer injuries in varying degrees of severity" (Nichols and Armstrong, 1973: 1). Moreover, they note that the scale of this—not to mention deaths recorded from industrial disease—is both "staggering" and "a sad commentary on the society in which we live."

How much, then, has changed in the past 40 years? On the face of it, a great deal. In 1975, Dawson, Willman, Bamford, and Clinton (1988: 225) used Health and Safety Commission and Executive (HSC/HSE) data to note that there were 620 fatal injuries and 328,500 nonfatal injuries across HSC/HSE-enforced workplaces. The HSE notes that the fatal injuries per 100,000 workers in 1974 were 2.9 (based on 651 fatalities). By 2008/2009, there were 179 such fatalities, at a rate of 0.6 per 100,000 workers.[1] That same year, there were 135,192 nonfatal injuries to workers (that is, major injuries and over-3-day injuries to workers, combined.[2] British workplaces may appear to have become far safer places; an observation that is sometimes held to vindicate the legislative and regulatory architecture established, following Robens, by the 1974 Act. However, such a claim is not so easily made on the basis of such data, and there are two fundamental, methodological caveats to note here regarding the comparability of data across time periods.

[1] http://www.hse.gov.uk/statistics/history/histfatals.xls
[2] See http://www.hse.gov.uk/statistics/history/histinj.xls

First, there are issues relating to recording injuries and deaths. HSE fatality figures generally include only deaths caused relatively quickly by sudden injury. Deaths that result from occupational disease and deaths in which there is a significant period of time between the injury and the death are only rarely recorded by the HSE in fatality statistics. This peculiarity alone ensures that the HSE headline figure of deaths takes account of only a small minority of occupational fatalities. Even the subcategory of fatalities that are recorded and reported in annual statistics by the HSE are gross underestimates. We have noted elsewhere that the real total of fatalities resulting from sudden injury recorded is around 3–4 times that recorded by the HSE (Tombs, 1999; Tombs and Whyte, 2008). What this means is that there is no way of knowing the full toll of deaths caused by work. If one includes within the occupational fatality totals those who are killed while driving as part of their work (Tombs, 1999), then, given the proliferation of motor vehicle use across the UK in the past 40 years, one might expect that hidden figure of deaths to be considerably greater as we write than was the case at the time of the Robens Report (1972).

Second, there are also problems related to nonfatal injury data. Due to widespread underreporting, HSE data significantly understates the actual level of work-related nonfatal injuries. The HSE itself documents the level of underestimate: for example, while there were 131,895 nonfatal injuries to employees reported under RIDDOR in 2008/2009 (502.2 per 100,000 employees), the 2009 Labour Force Survey found 246,000 reportable injuries occurred in the same year (at 870 per 100,000 workers). Indeed, it is well documented that each category of nonfatal injury data maintained under RIDDOR is subject to significant underreporting.

Now, underreporting was highlighted almost 40 years ago by the Robens Committee. It had noted, for example, an HMFI survey that "suggested that more than a quarter of accidents legally notifiable . . . are not in fact notified" (Robens, 1972: 135); that is, "less than 3 out of 4!" (Nichols and Armstrong, 1973: 2). Now, despite various and numerous attempts to improve reporting levels, and despite this being a legal requirement on employers, rates of reporting remain low, perhaps even as low as 30% (Davies, Kemp, and Frostick, 2007), a figure far worse than that cited by Robens, albeit perhaps a factor of improved methods for identifying such underreporting. In any case, this should temper the optimism regarding what appear to be dramatic downturns in the levels and rates of fatal and nonfatal injuries.

Turning to deaths due to occupational illness, the HSE does not "count" such deaths, rather it "estimates" them. The latest health and safety statistics (for 2010/2011) estimate that "the annual number of occupational cancer deaths in Great Britain is around 8000," adding that "results of work to estimate the number of cancers that result from current working conditions will be published in due course." It further estimates that about "15% of Chronic Obstructive Pulmonary Disease (COPD—including bronchitis and emphysema) may be work related," suggesting some 4,000 such deaths per annum "due to past occupational exposures to fumes, chemicals and dusts" (HSE, 2011a: 2). These estimates combine to produce a figure of some 12,000 deaths each year.

That said, even this figure itself looks certain to be an underestimate. One review of literature of estimates of occupational cancer in both the United States and the UK has concluded that the increasingly discredited methodology used by the HSE to reach its own estimates probably understates the incidences of such cancers by a factor of between two and four (Hazards, n.d.). Meanwhile, *Hazards* estimates that up to 20% of all heart-disease deaths have a work-related cause—for example, stress, long hours, shift work—which is about 20,000 a year. For all those diseases to which work can be a contributory cause, such as Parkinson's, Alzheimer's, motor-neurone disease, rheumatoid arthritis, chemical neurotoxicity, autoimmune conditions, and restrictive lung diseases, a further conservative estimate of about 6,000 deaths a year can be made. All of this adds up to a convincing (and lower end) estimate of deaths from work-related illness in the UK of up to 50,000 a year, or more than four times the official HSE figure.

This discussion of occupational diseases and the lack of our ability to know much about those diseases is closely related to the changing profile of work since 1973. If the decimation of the manufacturing industries and concomitant shift in the UK occupational structure appears to be a key factor in driving down sudden industrial *injuries*, the concentration of workers in clerical and service sectors also has consequences for their health. That is, the long-term shift from manufacturing to services is not one from relative danger to relative safety, as is often claimed. Indeed, the conditions that cause the bulk of occupational illnesses referred to above are either equally (or more) likely to arise in clerical and service industry workplaces as they are in manufacturing. Let us briefly sketch out some examples. First, asbestosis and mesothelioma are problems that are increasingly linked to people who work in the buildings that are full of asbestos: office workers, teachers and pupils in schools, and so on. If the first wave of asbestos-related fatalities fell disproportionately upon building workers, the victims of the current, unfolding wave have a very different profile. Second, stress-related diseases impact upon workers in office-based or service industry jobs no more or less than their counterparts in manufacturing occupations. Third, many of the manufacturing activities that have persisted or even emerged and thrived in the UK in the past 40 years have been characterized by high risks of dangerous occupational exposures: witness the emergent high rates of cancer and occupational health problems in the electronics industries and the various waste disposal and recycling industries. It is also, of course, of relevance that the levels of trade union organization tend to be particularly low among many of these new risk groups.

Of course, the above examples refer to specific, if highly significant, categories of occupational risk. Taken together, however, they indicate that, although the world of work has certainly changed, and the profile of the types of risks faced by workers has also changed, there is no reason to believe that the level of risk and the gross numbers of injuries, illnesses, and fatalities have significantly diminished since the publication of *Safety or Profit* (1973), if they have diminished at all.

We make these points as a cautionary introduction to our discussion of the new politics of regulation. For we would challenge any assumption that the basic political problem of worker exposure to hazards in the workplace has changed significantly. Work continues to kill, injure, and cut short people's lives through debilitating illnesses. What has changed, as we wish to document in the following sections, are the ways in which regulators represent their responses to these facts of capitalist life.

CONTEXTS FOR AND TRENDS IN LAW ENFORCEMENT UNDER NEW LABOUR

The analysis in this chapter draws upon an empirical analysis of trends in safety law enforcement during the period of the three Labour Governments, 1997–2010 (Tombs and Whyte, 2008, 2010a, 2010b). In this section, we present an overview of New Labour's policy approach to regulation in general, and then present some indications of enforcement based upon Health and Safety Executive data. In the following section, we consider the responses of the HSE to these parallel developments, focusing, in particular, on how it attempted to resolve the contradiction between government demands to reduce enforcement and its regulatory mandate to ensure the health, safety, and welfare of workers and members of the public who may be affected by work activities.

In 1997, New Labour began by renaming the Conservative's flagship Deregulation Unit as the Better Regulation Unit, with the Better Regulation Task Force established in the Cabinet Office. Regulatory Impact Assessments (RIAs) were introduced the following year. In 1999, the role of the Better Regulation Unit (renamed the Regulatory Impact Unit) was extended with a remit to ensure that RIAs were being implemented across government departments. RIAs aim to measure the costs and benefits of reforms on business, consumers, third-sector organizations, and public authorities of all proposed policy and legislative reforms. Yet they contain a structural bias toward less rather than more regulation in at least two ways. Their very rationale is the need to consider "the impact of any new regulations, before introducing them, to ensure any regulatory burden they add is kept to a minimum."[3] In other words, they are mechanisms that are based upon reducing regulation. Moreover, their economic form is likely to produce a financial argument for less rather than more controls on business activity, since the costs of meeting new regulatory requirements on the part of businesses are generally more calculable than are the economic or social benefits of such regulation (Cutler and James, 1996). In other words, they cement the business case at the heart of regulatory policy.

In April 2001, the Regulatory Reform Act allowed government ministers to order the reform of legislation with a view to removing or reducing regulatory "burdens." This law set the tone for New Labour's second period in office. In the same month, Tony Blair launched New Labour's manifesto for business before 100 corporate leaders in London. He committed Labour to develop, in its second

---

[3] *Scrutinising New Regulations*, at http://www.berr.gov. uk/whatwedo/bre/policy/scrutinising-new-regulations/page44076.html

term, a "deeper and intensified relationship" with business (Osler, 2002: 212). Commenting on Blair's pledges, Osler notes that "policies on offer that day included deregulation" (Osler 2002: 212). In the flush of Labour's second landslide victory in 2001, the gloves came off and a material and ideological assault on regulation was launched. The ideological assault took the form of a long-term, "drip-drip" type of discursive framing of regulation as "red tape." The impression conveyed was that regulation created a "burden on business." In 2004, Chancellor of the Exchequer Gordon Brown called for a break with the "old regulatory model" within which "everyone was inspected continuously, information demanded wholesale, and forms filled in at all times, the only barrier being a lack of regulatory resources." Launching what was known as the Hampton Review, he argued for a new model that he characterized as "not just a light touch, but a limited touch" (see Brown, 2005).

The Hampton reforms (Hampton, 2005) have at their heart a very carefully constructed rationale that defines regulation first and foremost in terms of its economic burden on business. Thus, Chapter 1 of the Regulatory Enforcement and Sanctions Act 2008 creates a remarkable new power for a Minister of the Crown to make an order that removes from government a "regulatory burden," defined in the Act as a "financial cost," an "administrative inconvenience," or "an obstacle to efficiency, productivity or profitability." There was by now a very unashamed and open honesty about the language being used; that is, in legislation and in policy, a very open admission that there is a direct relationship between the shift toward self-regulation and a neo-liberal profit-maximizing agenda. The explicit economic rationale at the heart of the Hampton reforms reached its high point in the new Regulators Compliance Code,[4] published in December 2007. This Code was introduced to address how "the few businesses" (para. 8) that break the law should be handled. In general, regulators, including the HSE, were advised that "By facilitating compliance through a positive and proactive approach, regulators can achieve higher compliance rates and reduce the need for reactive enforcement actions" (para. 8); they "should seek to reward those regulated entities that have consistently achieved good levels of compliance through positive incentives, including lighter inspections and less onerous reporting requirements" (para. 8.1); they should also "take account of the circumstances of small regulated entities, including any difficulties they may have in achieving compliance" (para. 8.1). If the rationale for these new realities of regulation was not clear enough, the document formalized the emerging conflict of interest for regulatory bodies when it emphasized that "regulators should recognise that a key element of their activity will be to allow, or even encourage, economic progress and only to intervene when there is a clear case for protection" (para. 3). Thus, the Hampton Review, and the reforms that followed, extended the scope and reach of the burdens on business agenda directly into the day-to-day work of inspectors, further marginalizing the enforcement role expected of regulators and giving renewed momentum to New Labour's pro-business trajectory.

---

[4] See http://www.berr.gov.uk/files/file45019.pdf

Hampton's reforms represent the culmination of two terms of New Labour's Better Regulation agenda. In this context, then, it is hardly surprising that every area of the HSE's formal enforcement activity declined significantly during the first decade of this century. Our documenting of regulatory activity across three Labour governments found that from 1999/2000 to 2008/2009, HSE inspections of workplaces fell by over two-thirds (from 75,272 to 23,004); there was a 63% decline in HSE investigations of incidents reported to it (from 11,462 to 4,272); a 29% fall in the number of all types of enforcement notices issued (from 11,340 to 8,079); and a fall in HSE prosecutions of 48% (from 1,616 to 837). Across this period, we have noted that those trends intensified during the period in which the Hampton Report was commissioned and published (2004 and 2005; Tombs and Whyte, 2010a, 2010b).

Of course, in many respects, those trends are part of a longer-term trajectory that stretches back at least as far as 1973. Nichols and Armstrong (1973) cited data from the 1972 Robens Report to note that in 1970, there were 300,000 factory inspections. By 1975, within one year of the passing of the HASAW Act, Dawson et al. (1988: 225) noted that there were 481,000 "visits" by all HSE enforcement agencies (excluding local authorities). As our data indicate, by 2008/2009, there were 23,004 inspections by the Field Operations Directorate (FOD), the largest section of the HSE and that which conducts the vast majority of inspections. While our figures for FOD inspections are not absolutely comparable with 1975 data for HSE visits—probably about a third of the latter, some 150,000, would take the form of an inspection (*Hansard*, HC Deb 14 April 1976 vol 909 cc629-30W)— we can still say with some certainty that the annual number of workplace inspections conducted today is a fraction of that in 1975.

As our data above indicates, there has also been a long-term decline in prosecutions; and those prosecutions appear to have been replaced by enforcement notices (indeed, within the data, it is clear that the most serious forms of enforcement notice—the Prohibition notice—has fallen much more sharply than the least serious, the Improvement notice). By way of comparison in terms of absolute levels of prosecutions, Nichols and Armstrong (1973: 2) cited data from the Robens Report (1972) to note that, in 1971, there were 1,330 prosecutions. By 1975, within one year of the passing of the HASAW Act, Dawson et al. (1988: 228) noted that there were 1,588 HSC/HSE prosecutions (while the conviction rate was not available, it is almost certain that some 90% of the prosecutions would have been successful). As our data indicate, by 2008/2009, there were 837 prosecutions instituted by the HSE—a dramatic decline in the most serious form of enforcement action. Put simply, from 1975 to 2008/2009, prosecutions as well as inspections have declined significantly.

Now, those who are familiar with the Robens Report (1972) and with Nichols and Armstrong's 1973 analysis of it might legitimately see the current enforcement levels not only as a function of "regulatory surrender" (Tombs and Whyte, 2010a) but as being consistent with Robens' own views of law and enforcement. As the Robens Report noted,

104 / SAFETY OR PROFIT?

> The process of prosecution and punishment by the criminal courts is largely an irrelevancy. The real need is for a constructive means of ensuring that practical improvements are made and preventative measures adopted. Whatever the value of the threat of prosecution, that actual process of prosecution makes little direct contribution towards the end. On the contrary, the laborious work of preparing prosecutions . . . consumes much valuable time which inspectorates are naturally reluctant to devote to such little purpose (para. 261).

The logical conclusion that follows from the common interest principle is decriminalization; to use the criminal law against employers is simply not necessary and indeed counterproductive when improvements in safety management are sought.

More generally, as Nichols and Armstrong noted, the common interest principle was the premise upon which a shift toward "deregulation" was to be based:

> If apathy were the problem, the recommendation that the amount of external regulation be reduced might have been the answer. But if, as our examples indicate, men are taking risks as a response to the pressure for production, reducing the amount of regulation would only allow freer rein to that pressure (Nichols and Armstrong, 1973: 21).

The common interest principle was also the premise that justified the establishment of a tripartite institutional structure of HSC/E that gave trade unions and businesses representation on the board overseeing the work of the regulator. In order to explore how this system of institutionalized "common interest" operates empirically, the following section presents an analysis of the way that debates on enforcement have been conducted in the Health and Safety Commission.

THE FAILURE OF TRIPARTISM

The principle of a natural identity of interests was extended to the institutional governance of the Health and Safety Executive. The HSE's governing body, the Health and Safety Board (formerly the Health and Safety Commission)[5] is structured at board level to incorporate representatives of business, government, and the trade unions. A large number of advisory groups in different industrial sectors that provide support to the board are also structured on tripartite principles.

The trends that we refer to briefly above are indicative of some undeniably dramatic changes in the law enforcement activities of the HSE. The way that the tripartite system has responded to those trends therefore potentially represents an interesting case study in the way that the "common interest" principle plays out institutionally in the Health and Safety Commission. To this end, this section

---

[5] Following the merger of the Health and Safety Commission and Health and Safety Executive on April 1, 2008, the Health and Safety Commission became the HSE Board. When we refer to meetings/papers here that were tabled/discussed prior to April 1, 2008, we refer to these as relating to them emanating from the Commission (HSC).

summarizes HSC discussions of the sharp declines in inspections, investigations, and enforcement action. The following paragraphs present an analysis of the minutes of Health and Safety Commission meetings and (following the merger of the Health and Safety Commission with the Health and Safety Executive in April 2008) the minutes of the Health and Safety Executive Board.

Before we develop this analysis, it is worth noting three points about the way in which those minutes are recorded and the form in which they are presented. First, they are a brief summary of what are very often lengthy discussions. For this reason, they cannot be interpreted as a detailed account or record of the Commission meetings; they merely highlight the most significant aspects of Commission discussions. Second, with the exception of the Chair of the HSC, the minutes generally avoid naming individual commissioners or attributing names to particular views. Thus, it is difficult from reading the minutes to clearly interpret the fault lines or the divisions of opinion across constituent members. Third, and following the previous point, the position of the Commission is almost always expressed in the minutes as a consensus. The minutes very rarely identify conflicts of opinion or disagreements between commissioners. When the minutes do appear to be expressing disagreements, they are expressed as a "wide ranging discussion" (see, for example, Minutes, 6th April 2004: para. 3.3) or noted as "a number of concerns" about a particular issue (see, for example, Minutes, 14th March 2006: para. 9.2). The over-riding impression given by the minutes of those meetings is that there is broad agreement across commissioners on most issues. The way that this consensus is recorded may or may not reflect accurately the deliberations and discussions at those meetings, but it is highly significant that the minutes, as a *public record* of the Commission's business, represent the view of the Commission in a consensual way. It is this apparent consensus, as expressed on public record, and the impact that those publically expressed deliberations and discussions had on debates on enforcement between January 2003 and December 2010 that the following analysis explores in a little more detail. We have four observations to make.

The first thing that is notable about the minutes of the meetings throughout the period discussed is the lack of discussion of enforcement. Enforcement is not a standing item at those meetings but is generally discussed only in relation to policy papers presented to the Commission. Mention of the collapse in enforcement that we describe in this chapter warrants significant discussion in only a handful of meetings over the 8-year period analyzed.

Second, when enforcement *is* discussed at those meetings, commissioners generally do not take the opportunity to question those downward trends. Very often, discussions of enforcement, particularly in relation to FOD, pass without any comment on trends over time at all (see for example, Minutes, 9th January 2007: para. 6.2). Even when the discussion is explicitly focused on the centrality of enforcement to strategy, as in discussions on the Regulatory Decision Making Audit (HSE/09/33), or Revised Guidance to Enforcing Authorities (HSC/08/12), there is no discussion about the collapse in enforcement (Minutes, 29th April 2009; Minutes, 12th February 2008). Moreover, when commissioners discussed particular cases in which the HSE's enforcement strategy had been questioned publically, they similarly did not take the opportunity to discuss enforcement policy or strategy. Thus, for example, it was noted by the Chief Executive in the September 2007 meeting that that there had been an independent report on the ICL Plastics Explosion in Glasgow in May 2005. The report had publically criticized the HSE's enforcement strategy, as had several national newspaper reports. However, the Commission dismissed those concerns without discussion, recording the following statement: "There had been unjustified and unsubstantiated criticism of the HSE which should be refuted" (Minutes, 4th September 2007: para. 2.2).

Third, we find a similar lack of acknowledgment of those downward trends in discussions of the Hampton Review (2005). The position expressed by the Commission is generally one of enthusiastic support for a general shift in emphasis from enforcement to "encouraging" duty holders to comply (see, for example, Minutes, 14th October 2003: para. 5.3; Minutes, 9th December 2003: para. 3.2), and for the principles underpinning Hampton (see, for example, Minutes 3rd August, 2004: para. 3.2), even recognizing it as a "vote of confidence in HSC/E" (Minutes, 5th April, 2005: paras. 7.2, 7.3). The Commission is equally supportive of some of the most business-friendly principles that underpin the Hampton Report. Thus, when it came to implement the new Regulator's Compliance Code (see discussion above), the Commission recorded its support for the Code, since it "did not undermine either the HSE's and LA's statutory duties to enforce HASAW Act or the regulatory approach" (Minutes, 17th July 2007: para. 5.3). A sentiment of virtually unqualified support for the Hampton agenda in the Commission is apparent, if rather surprising, given the implications of the agenda for reasserting business interests above those of the organized workforce. In relation to the specific enforcement issues dealt with in the Hampton Report, there is little discernible difference in this position.

Fourth, on the rare occasions that trends in enforcement *were* discussed, the trends remain unproblematized. Some discussions of enforcement trends were couched more vaguely in terms of the balance in FOD inspectors' activities between "proactive" or "reactive" work (see, for example, Minutes, 11th February 2003; Minutes, 11th October 2005).[6] In the only recorded discussion of the year-on-year decline of enforcement indicators, the Commission is moved to congratulate the HSE for apparently "increasing enforcement" and using those figures to offset criticism of the enforcement record: "The Commission was pleased that the enforcement action had increased. This should be publicised because there had been criticism in the media of the spending review and the impact of cuts on enforcement" (Minutes, 15th May 2007: para. 5.2). Thus, instead of using this as an opportunity to explore or question the long-term decline in enforcement, the Commission advocates the selective release of these figures for a particular year for public relations purposes.

---

[6] In this context, reactive work includes investigations of RIDDOR injuries and incidents as well as enforcement work.

Taken together, those minuted discussions indicate that the Commission generally failed to recognize the long-term decline in inspection, investigation, and enforcement as a problem worthy of consideration in any great detail, save for this decline to be glossed over or refuted.

The trends to which we have already devoted some attention (a two-thirds fall in inspections and investigations and an almost halving of prosecutions) continued throughout the post-2003 period and effectively went unchecked by the HSC and the HSE Board. Indeed, the Board actually signalled its clear support for FOD's enforcement approach (Minutes, 30th June 2010: para. 5.3).

What all of this indicates is that, for the purposes of board-level, minuted, discussions, a consensus around the "better regulation" agenda has ensured that even the most dramatic downward trends in enforcement have passed without serious challenge.

SAFETY AND PROFIT

In one sense, what is implied in the Better Regulation consensus that was forged by New Labour—and which was both championed within and further consolidated by the HSC/E—is a confirmation of Nichols and Armstrong's (1973) analysis: that regulation is explicitly characterized as a financial burden and therefore is a threat to economic progress, success, and so on. In other words, the regulation of safety *is* a threat to profits.

In another sense, there is also a fundamental contradiction that has to be dealt with at the level of policy: how to secure safety *and* profit when the incompatibility of those aims is revealed by real cases of occupational fatalities. In other words, common interest claims are always likely to be vulnerable to challenge.

The Better Regulation agenda sought to resolve this contradiction in two intimately related ways: first, through a theoretical understanding of the nature of business which has close similarities to the "identity of common interests" philosophy that supported new forms of self-regulation; second, through the pseudo-science of targeted intervention. We now briefly consider each in turn.

For New Labour, a key assumption is not only that businesses must be encouraged to be "moral" and "responsible," but that most *are* moral and responsible: it is only the minority that need to be monitored. Thus, the Hampton agenda enthusiastically endorses twin-track regulation whereby regulatory interventions are "targeted at the worst offenders" (Hampton, 2005: 26). It is an approach underpinned by a blind faith in corporate morality that enables the majority of good corporate citizens to be left to their own devices.

Most fundamentally, corporate social responsibility (CSR) involves the claim that businesses seek to respond to more general concerns, values, or pressures in society; that is, to take on commitments over and above those placed upon them by legal duties. In other words, it has recently been claimed that CSR

> involves a shift in the focus of corporate responsibility from profit maximisation for shareholders within the obligations of law to a responsibility to a broader range of stakeholders, including communal concerns such as protection of the environment, and accountability on ethical as well as legal obligations (McBarnet, 2007: 9).

This entails a shift from "'bottom line' to 'triple bottom line'" (McBarnet, 2007: 9) of economic, social, and environmental performance. Companies should operate in ways that secure long-term economic performance by avoiding short-term behavior that is socially detrimental or environmentally wasteful. The principle works best for issues that coincide with a company's economic or regulatory interests (Porter and Kramer, 2006: 82). For Porter and Kramer, "corporate success and social welfare" are not a zero-sum game (2006: 80). From this perspective, CSR is neither reactive nor defensive but strategic and proactive, part of a classic win-win game: "Successful corporations need a healthy society. . . . At the same time, a healthy society needs successful companies" (p. 83). Social health is thereby equated with corporate success.

The HSE's own "version" of social responsibility is based upon exactly the same theory of CSR—summed up in the phrase "win-win." As the HSE Deputy Director General Jonathan Rees noted, "Sensible health and safety management is a key part of effective business management" so that the HSE seeks to highlight "the vital contribution that such an approach can have on the performance of businesses as well as on employees' welfare: a true win win" (HSE, 2004). The HSE's own statement on Corporate Social Responsibility (HSE, n.d.) reveals that the HSE's mission "to ensure that the risks to health and safety of workers are properly controlled" is pursued through encouraging organizations to "improve management systems to reduce injuries and ill health." These systems—this "effective management of health and safety"—are thereby "vital to employee well-being"; have "a role to play in enhancing the reputation of businesses and helping them achieve high-performance teams; and are "financially beneficial to business."

And here we reach the crux of the matter for the HSE's contemporary approach to "regulating" business: there is a business case for better safety management, which rational managements can identify and operationalize. This requires a concept of active management or leadership, on which the HSE has their own positions, as set out in its longstanding *Successful Health and Safety Management* (HSE, 1997) and in a much more recent document, *Leading Health and Safety at Work* (Institute of Directors and Health and Safety Executive, 2009). This, in turn, obviates the need for external enforcement, that is, law reform or enforcement of existing law. Thus, for example, the HSE document cited above, *Leading Health and Safety at Work*, acted as a bulwark against the campaign for legal duties on directors. It is also significant that this document was jointly authored by the HSE and the Institute of Directors. It is a voluntarism which simply flies in the face of evidence regarding "what works" (Davis, 2004).

This faith in CSR in general, and in responsible business leadership in particular, thus leaves the problem of those businesses who are not law abiding. Thus, a key focus of the Hampton agenda is upon rogue offenders, that is, "persistent offenders" that "deliberately flout the law and undercut honest business" (Hughes, 2007: 6). This is, of course, perfectly in tune with the dominant academic approach to regulatory enforcement. For example, in the classic compliance-oriented enforcement text, *Going by the Book*, in an attempt to rebut the argument that all corporations are amoral calculators, Bardach and Kagan assume, "for analytical purposes," that at most, what they cutely term "bad apples" in fact "make up about 20% of the average population of regulated enterprises in most regulatory programs" (Bardach and Kagan, 1982: 65).

Herein enters targeted intervention, or risk-based enforcement, which allows more to be done with fewer resources and in the context of less apparent overall enforcement activity, since the majority of businesses are likely autonomously to comply when faced with a combination of persuasion and market incentives. While risk-based regulation is often narrowly cast as a method whereby scarce inspection resources are allocated, it has a wider and increasingly significant impact upon regulators. It helps regulators to "structure choices across a range of different types of intervention activities, including education and advice" (Black, 2010: 186). In other words, the complex and often convoluted logic of risk-based regulation provides a rationale for a shift toward more consensus or compliance-based strategies, which appeal to the cooperation and good will of business.

Notwithstanding the importance of risk techniques for providing basic information about duty holders, "risk" as a concept is used as a politically convenient concept that opens up space for shifting the terrain of regulatory intervention (Tombs and Whyte, 2006). Intervention strategies based upon risk factors or risk indicators are the technical means for achieving earned autonomy. The production of risk indicators enables regulatory interventions to be targeted at the worst offenders. In this sense, they are the technical means to achieving a CSR-based targeted intervention approach.

CONCLUSION

Nichols and Armstrong's 1973 analysis took place on the cusp of a major ideological shift in late 20th century capitalism: a shift that witnessed the rise to dominance of a set of ideas that have been generally described as "neo-liberalism." Yet, for that seismic political-economic shift, the perspicacity of their analysis for understanding the mystifying "common interest" remained.

Things *have* changed significantly at the levels of the social relations of production and the wider balance of class forces, as well as, more particularly, at the level of ideological production. The common interest claims necessary to sustain a tripartite system of health and safety regulation remain, while the key ideological support of these has been provided by the ideological delusion of corporate social responsibility. It is this ideological architecture that both smoothed the way for, and was in turn bolstered by, the New Labour/Hampton Better Regulation agenda in general, and its new, targeted intervention strategies in particular. The result has been a collapse in regulatory intervention; in other words, Robens' (1972) original plan to replace external regulation with self-regulation has come to pass.

Perhaps what was not envisaged by Robens in 1972 is the extent to which the tripartite system has failed to ensure that worker interests are represented at the level of policy. We have explored one aspect of this in our analysis of the business of HSC. It is worth noting a supplementary point that we did not include in our earlier analysis: at no point in 8 years of HSC meetings is there any record of the business case for safety being challenged, or even discussed critically. It became an assumption, one embedded in the institutional architecture of the health and safety system.

This apparent lack of challenge in turn raises questions over the effectiveness, relevance, or legitimacy of TUC representation on the Commission. Either TUC representation has not challenged the HSC's policy trajectory or, if it has, then such challenge has not been placed on the formal record. Whichever is the case, we must ask some pressing questions about the extent to which the trade unions have been—wittingly or unwittingly—incorporated into an agenda that has significantly disempowered workers in struggles for improved health and safety conditions. If tripartism grants the HSC legitimacy to pursue a path toward regulatory impotence, we must ask what that tripartism actually means and whether it is actually worth defending.

HSC/E's elegant and creative embracing of the better regulation agenda has done nothing more than accelerate its own decline. If HSC/E believed that in acquiescing to the Better Regulation agenda as developed under New Labour it was preemptively securing its role and function, it has quickly and clearly been shown to be mistaken. In shifting, even helping to shape, the Better Regulation agenda rather than openly challenging it, HSC has left itself even more exposed to further attack; for regulation can always be "better," as defined within that agenda.

This much was clear when, in the run-up to the General Election of 2010, even after 10 years of regulatory degradation, the consensual view of all three main parties was that health and safety was still a site of overregulation. Some 5 weeks after the coalition was formed, Lord Young was appointed to examine health and safety law and "the compensation culture." His report, *Common Sense, Common Safety*, was published in October 2010, with a series of recommendations aimed "to free businesses from unnecessary bureaucratic burdens" (Young, 2010: 9). These were all accepted by HSC (2010). In the same month as Young's review was published, the Department of Work and Pensions (DWP) announced, as a result of the government's spending review, that funding for the HSE would fall by 35% by 2015. In March 2011, the DWP launched its *Good Health and Safety, Good for Everyone* document, designed (as its subtitle states) to set out "The next steps in the Government's plans for reform of the health and safety system in Britain." One of these steps was that the HSE would "reduce its proactive inspections by one-third (around 11,000 inspections per year)" (DWP, 2011: 9), by ending proactive inspections to "lower risk areas" (DWP, 2011: 9). The same report established a *further*, formal critical examination of health and safety law: the Lofstedt Review was to "consider the opportunities for reducing the burden of health and safety legislation on UK businesses" (DWP, 2011: 2).

HSC/E has been institutionally undermined by a repackaged business case and has been completely complicit in that process. Yet this has not insulated it from further, sustained attack. For if the business case was, as we have argued, based upon claims for profits *and* safety, the opposite is now being proposed—that we can have safety *or* profits; that "we" must accept less regulation for financial recovery. Thus, worker protection beyond a minimum has become an unaffordable luxury in the age of austerity. The new version of the safety/profits dichotomy that Nichols and Armstrong (1973) so brilliantly exposed almost 40 years ago is increasingly positioned openly as a choice between safety *or* profits; that "we" must accept less regulation for financial recovery. It is this feature of the contemporary British economy that, if anything, means the ability of workers to protect themselves at work has become more precarious and uncertain than in 1973.

http://dx.doi.org/10.2190/SOPC6

CHAPTER 6

Decriminalization of Health and Safety at Work in Australia

Richard Johnstone

INTRODUCTION

This chapter is inspired by the critique, in Nichols and Armstrong's pamphlet *Safety or Profit* (1997: ch. 3), of the British Robens Report's (1972) conception of work health and safety enforcement, and by Nichols and Armstrong's exhortation to see work-related injury, disease, and death "in their total situation" and "in the context of the *social relations of production*" (emphasis in original) and for greater "legal restraint" and stronger enforcement of health and safety legislation. I was particularly struck by the pamphlet's critique of the Robens Report's rather misguided suggestion that "accidents" are "events" and that they are "rare" or "isolated."

This chapter analyzes the approach to enforcement of Australian work health and safety regulatory agencies. On one level, it would appear that there has been a renewed interest in stronger enforcement—including more robust use of prosecution—of the Australian work health and safety statutes since the early 1980s. Certainly, as this chapter will show, the maximum financial penalties for work health and safety offenses have increased dramatically in that period, prosecution is more prominent, and in some cases, the fines imposed by the courts have been substantial. The chapter argues, however, that despite the rhetoric of stronger enforcement and more robust prosecution, the dominant ideology of work health and safety enforcement—ambivalence about whether work health and safety offenses are "really criminal" and viewing prosecution as a "last resort" in the enforcement armory—still dominates the approach of Australian work health and safety regulators. Drawing on empirical research into work health and safety prosecutions in the 1980s and 1990s, the chapter also argues that, because prosecution is usually focused on an "event" that resulted in injury or death, the way in which prosecution is used in Australia, and the legal structure surrounding prosecution, plays a further role in "decriminalizing" work health and safety offenses.

114 / SAFETY OR PROFIT?

The chapter begins with a brief reminder of how advice and persuasion became the dominant approach to enforcement endorsed so uncritically by the Robens Report (1972). The chapter then reports on research into work health and safety prosecutions, and then critically examines post-Robens debates and developments in work health and safety enforcement in Australia since the early 1980s.

"HOSTAGES TO HISTORY"

Until very recently, the Australian approach to work health and safety regulation has drawn heavily upon the UK regulatory model. The first Australian Factories and Shops Acts (Victoria, 1973 and 1885; South Australia, 1894; NSW, 1896; Queensland, 1896; Western Australia, 1904; and Tasmania, 1910) were based upon the UK Factories Acts then in force; indeed, some of the Australian statutes were largely a "cut and paste" of the prevailing UK provisions (see Johnstone, 2000, 2004a: 42; 2009). Similarly, the model of work health and safety regulation proposed by the Robens Report (1972) had a major influence on the shape of the Australian work health and safety statutes introduced from the early 1970s (see Johnstone, 2004a: 80–84). The UK influence was not confined to legal structure and content: from 1873, the Australian work health and safety inspectorates largely adopted the UK approach to enforcement of the work health and safety statutes.

Soon after the formation of the UK factories inspectorate in the Factory Regulation Act of 1833, there emerged an enforcement culture that focused on securing compliance with the Act through advice, persuasion, and negotiation rather than prosecuting contraventions (what could be called the punitive or deterrent approach); prosecutions were reserved for "serious" or "wilful" offenses. For example, Bartrip and Fenn (1983: 206) describe

> an emphasis on persuasion, in all its varieties, rather than on prosecution following a detected contravention against the Act. Prosecutions were reserved for "serious" or "wilful" offences, while most detected offenders were given the opportunity subsequently to comply with the law and therefore escape liability.

Bartrip and Fenn (1983) argued that the inspectors adopted this approach because they had limited resources, and resort to persuasion and advice was the most cost-effective enforcement technique. Carson (1979, 1980), from a more critical political economy perspective, showed that by 1833, a number of large, urban manufacturers, for various reasons, supported the provisions of the 1833 Act, including strong enforcement by the inspectorate, but that the inspectorate by the late 1840s had adopted an advise and persuade approach as a way of coping with the clash between the movement toward effective regulation and the fact that contraventions were widespread. As Carson vividly put it, heavy use of prosecution would have entailed "collective criminalisation" of employers "of considerable status, social respectability and . . . growing political influence" (Carson, 1979: 48). Carson showed that the Factories Acts manifested the contradictory tendencies of their own development and institutionalized the "ambiguity" of factory crime so that it was not seen as "really criminal" (Carson, 1980); it was frequently breached and substantially tolerated in practice (i.e., "conventionalized") (Carson, 1979). This was reinforced by the fact that, at the initiative of inspectors in order to facilitate enforcement, from 1844 the Factories Acts generally imposed strict liability (meaning that liability did not require proof of negligence or intention to harm). A further factor tending to undermine the "criminality" of work health and safety offenses is that since the early Factory Acts, most offenses have been "inchoate," in that they can be committed when work conditions put workers at risk of injury, disease, or death regardless of whether or not injury, disease, or death actually eventuates.

It is important to remember that this approach to enforcement, dominant since the late 1840s, is historically contingent, the result of the particular historical circumstances facing the early UK factory inspectors. Nevertheless, this differentiation of work health and safety crime from "real" crime has been widely accepted by researchers, regulators, lawyers, and the community in Australia and in the UK, and it suggests that the ideologies of the ambiguity and conventionalization of work health and safety crime have become deeply embedded in the Australian and UK discourse about work health and safety enforcement. One of the many reasons for this is that, as the Nichols and Armstrong (Nichols, 1997: ch. 3) pamphlet reminds us, the advise and persuade approach and an underresourced inspectorate were strongly endorsed by the Robens Committee.

THE AUSTRALIAN APPROACH TO WORK HEALTH AND SAFETY ENFORCEMENT

It would appear that the Australian inspectorates immediately adopted the advise and persuade approach to work health and safety enforcement. For example, the Chief Inspector of the Queensland Inspectorate wrote (Annual Report, 1897: 6; Maconachie, 1986: 81) that inspectors sought to "secure compliance with the provisions of the Act . . . without having to recommend stronger measures than persuasion." The Chief Inspector of the Victorian Factory and Shops inspectorate in the period 1962 to 1973 wrote that (Prior, 1985: 54)

> most inspectorates . . . see as a failure any inspector who constantly has to launch prosecutions in order to obtain compliance. They see the legislation they administer as being remedial rather than punitive in nature, i.e. they are there to improve the conditions of work, not to make the employer or employee suffer penalties for breaches of the law.

This approach to enforcement was still the norm during the last decade. During this period, the nine generalist Australian work health and safety inspectorates averaged about one inspector per 10,000 employees (Workplace Relations Ministers' Council, 2008, 2010), or between 1,000 and 2,300 workplaces per inspector (Productivity Commission, 2010: 107, Table 5.5). This suggests that it was highly unlikely

116 / SAFETY OR PROFIT?

that there were enough inspectors to visit all workplaces over the cycle of a few years, let alone make frequent inspections to promote compliance. The Productivity Commission's (2010: 115, Table 5.9) data for 2008/2009 showed that the number of inspections made by each inspectorate varied considerably, as Table 6.1 shows.

When inspectors did take enforcement action, it was mostly to advise or educate duty holders. Table 6.2 shows that when formal enforcement action was taken, the dominant method was to use administrative sanctions, with the improvement notice (requiring a duty holder to remedy a specific contravention of the work health and safety Act within a specified time) used far more than the prohibition notice (directing that an activity stop until an immediate risk to health and safety was mitigated) or the infringement notice (on-the-spot fine; see further the discussion below).

Table 6.2 shows that prosecution was taken relatively infrequently, and the prosecution data for the state initiating the most prosecutions, New South Wales, clearly shows that the prosecution rate, having risen significantly in the 1980s, was falling markedly. Although the work health and safety statutes impose absolute liability offenses, qualified by "reasonably practicable," and therefore do not require prosecutors to prove intention of any kind on the part of the defendant firm, prosecutions generally are taken only because the inspectorate considers that the defendant has been "morally blameworthy" in its contravention of the work health and safety legislation (see Johnstone 2003: chs. 3, 4). In sum, although it would appear to most observers of work health and safety regulations in Australia that regulators have taken a far more robust approach to enforcement since the early 1980s, including a greater use of prosecution, this section of the chapter shows clearly that, in fact, work health and safety offenses have been conventionalized, that is, rarely prosecuted.

Research on Work Health and Safety Prosecutions

A major empirical study of 200 work health and safety prosecutions in Victoria, Australia, from 1986 to 1998 showed how work health and safety offenses are "decriminalized" in another sense when a prosecution is taken (Johnstone, 2003). One of the findings of the study was that the average fine imposed by magistrates' courts in successful prosecutions of work health and safety offenses during the period

Table 6.1 Total Number of Workplace Inspections Made by
Work Health and Safety Inspectors in 2008/2009, Australia States

Activity	Cwith	NSW	Vic	Qld	SA	WA	Tas	NT	ACT
Total inspections	580	13,452	42,169	16,852	19,934	11,339	6,280	4,007	2,304

Source: Productivity Commission, 2010; 115, Table 5.9 (see original table for footnotes qualifying the data further).

of the study was 21.6% of the maximum fine available for each offense prosecuted. Stated in absolute terms, the average fine in 1983 was just over A$294 (14.45% of the possible maximum fine); in 1986, just over A$402 (20.94%); in 1990, five years after a new Occupational Health and Safety Act 1985 came into force with much higher maximum penalties, A$2836 (22.92%); in 1994, just over A$7,808 (22.4%); in 1997, just over A$7,954 (21.1%); and in 1999, just over A$14, 673 (26.7%). These are not fines that would be likely to have much of a deterrent effect in anything other than a small firm.

The study suggests that one explanation for the low penalties had to do with the "event focus" of prosecution and the ease with which the defendant's legal representatives could use this event focus to decontextualize the offense and make it appear far less serious than it actually was when presenting arguments in mitigation of penalty. I noted earlier in this chapter that work health and safety offenses are inchoate in the sense that a contravention can occur before an injury or fatality occurs. Nevertheless, the vast majority of work health and safety prosecutions in Australian courts focus on an event—a work-related incident leading to an injury or death. Indeed, the prosecution policies of all of the work health and safety regulators specified that the usual trigger for a prosecution would be when an injury or fatality had taken place. The study of prosecutions in Victoria (Johnstone, 2003: 91–93) found that 87% of prosecutions were initiated after an injury or fatality at work, and there were similar findings in a 1996 study in New South Wales (see Gunningham, Johnstone, and Rozen, 1996). One of the consequences of this event focus of prosecution is that focusing on a specific incident "splinters" the event from its broader context (see Johnstone, 2003: 208; Mathiesen, 1981: 63) and focuses the court's attention on the incident and away from the underlying system of work, the defendant's approach to the systematic management of work health and safety, and broader features such as production pressures that override health and safety concerns. This is exacerbated when the prosecutor is forced by the rules of criminal procedure (e.g., the "rule against duplicity"; see Johnstone, 2003: 110–112) to issue a number of charges focusing on different aspects of the single event, such as training, instruction, supervision, and the work system.

Once the event is drawn out of its context, the defendant, in mitigation of penalty, can keep the court's attention on the minute details of the event and away from broader structural concerns. The study found that defense counsel used a number of very common arguments to further "isolate" the event from its work health and safety context.

The first technique was to use the detailed scrutiny of the event to shift the blame for the event onto another person; most commonly the injured worker (see Johnstone, 2003: 211–215), but also onto work health and safety inspectors who had "missed" the contravention in their previous inspections, or onto the seller of plant who had supplied the defective plant to the defendant.

A second technique was to argue that the defendant was a "good corporate citizen" with an unblemished record and a good attitude toward work health and safety (which occasionally included reference to the good character of individuals

Table 6.2 Enforcement Action Taken by Work Health and Safety Inspectorates in Australia, 2001–2010 by Jurisdiction

	Year	NSW	VIC	QLD	WA	SA	TAS	NT	ACT	Aust Gov	Total Aus
Number of infringement notices issued	2001–02	1,471	n/a	99	n/a	n/a	n/a	71	0	n/a	1,641
	2002–03	1,289	n/a	289	n/a	n/a	n/a	242	0	n/a	1,820
	2003–04	915	n/a	488	n/a	n/a	n/a	31	0	n/a	1,434
	2004–05	1,652	n/a	462	n/a	n/a	n/a	7	8	n/a	2,130
	2005–06	1,195	n/a	499	n/a	n/a	n/a	47	28	n/a	1,769
	2006–07	726	n/a	612	n/a	n/a	n/a	173	8	n/a	1,519
	2007–08	620	n/a	643	n/a	n/a	37	201	13	n/a	1,514
	2008–09	686	n/a	506	n/a	n/a	49	0	10	n/a	1,251
	2009–10	688	n/a	393	n/a	n/a	56	0	6	n/a	1,143
Number of improvement notices issued	2001–02	10,517	11,922	6,246	9,818	1,025	420	19	77	8	40,065
	2002–03	12,646	14,964	11,136	10,263	1,977	346	22	80	18	51,452
	2003–04	17,927	12,492	16,200	11,848	2,743	198	29	202	17	61,662
	2004–05	18,213	12,117	13,348	12,391	4,688	423	17	163	12	61,381
	2005–06	14,832	11,168	16,463	11,891	3,573	297	49	427	12	58,517
	2006–07	13,243	12,040	14,631	10,249	3,258	188	30	137	37	53,830
	2007–08	13,109	10,279	14,390	9,724	2,328	161	136	129	18	50,290
	2008–09	10,832	18,363	8,149	9,833	2,396	169	209	99	31	50,081
	2009–10	12,161	21,600	9,057	10,640	1,841	224	132	187	36	55,898

Number of prohibition notices issued

Year										Total
2001–02	786	3,102	1,188	887	191	109	25	39	2	6,331
2002–03	779	2,904	1,256	895	364	131	56	48	9	6,444
2003–04	1,139	2,308	1,886	870	814	87	14	50	6	7,020
2004–05	1,421	2,308	1,788	963	899	266	14	66	20	7,751
2005–06	1,212	1,876	2,223	708	623	125	54	88	10	6,981
2006–07	1,127	1,538	1,434	629	732	105	65	57	6	6,697
2007–08	994	1,043	2,784	676	588	113	61	94	19	6,375
2008–09	767	1,078	2,278	721	630	112	69	101	16	5,776
2009–10	856	928	2,277	705	628	167	51	103	26	5,744

Number of prosecutions resulting in conviction

Year										Total
2001–02	455	115	114	41	8	11	2	0	0	746
2002–03	443	105	101	38	22	24	0	2	0	735
2003–04	399	110	120	43	30	7	0	5	0	714
2004–05	384	93	156	48	31	7	0	11	0	731
2005–06	340	70	143	41	51	12	0	5	0	662
2006–07	300	87	102	29	56	16	2	2	1	595
2007–08	182	107	83	23	51	18	10	4	0	479
2008–09	96	107	102	18	62	6	5	3	2	402
2009–10	na	na	na	na	na	na	na	na	na	na

Source: Workplace Relations Ministers' Council (2008–2010).

n/a: Not applicable; na: not available.

Note: These figures do not record the fact that in Queensland, a great proportion of prosecutions over the past decade have resulted in fines without conviction. This is discussed briefly near the end of this chapter.

running the firm or the firm's contribution to the community). This was a very difficult plea for the prosecutor to challenge because the prosecutor, traditionally, has a very limited role in the sentencing process and, in the cases in the study, was very reluctant to challenge assertions made by defense counsel. In any event, most prosecutors did not go to court armed with detailed information about the defendant's previous work health and safety record and attitude.

The third technique was to "individualize" the event (Mathiesen, 1981: 58), and to argue that the event was a "freak accident" or "one off" event, and that the exceptional, the unforeseeable, and hence unpreventable had occurred. Most incidents, when examined in great detail, have unique features and often occur because of a coincidence of factors. They can appear to be a freak accident if the analysis is not conducted at the level of whether the system of work was safe and without risks to health, and whether that safe system was properly implemented and enforced. For example, if chemicals for use in a smelter are stored in unlabeled bags together with other chemicals in unlabeled bags, the defendant might focus on a series of coincidental events and argue that "evil chance" resulted in the wrong bag of chemicals being used (Johnstone, 2003: 222–223); whereas, if the analysis is at the level of the system, it is clear that the fault lay with inadequate labeling and storage systems. To borrow the terminology of the Nichols and Armstrong pamphlet (1997: 51), the prosecution process enabled defense counsel to construe issues as isolated events when proper analysis would have portrayed them as "process failures."

Used together, the second and third techniques suggest that the defendant has an exemplary history and approach to work health and safety, and that, in the particular case before the court, it had been unfortunate, and an incident had occurred because of an unpredictable confluence of circumstances.

A fourth technique was to ensure that the event was isolated in the outmoded past (Mathiesen, 1981: 68) and to show that since the event, things had changed: a new management team had been introduced, the plant or the system of work had been improved, and so on. The consequence, it was argued or implied, was that the court had no cause to deter, rehabilitate, or punish the defendant because the wrong had been corrected.

The point of these techniques (see Johnstone, 2003: ch. 7) is that attention is drawn away from the defendant's approach to work health and safety management and performance, and from "the total situation," and focused on the event, which is then shown to be a one-off incident, for which someone else (and not the exemplary employer) was to blame, and which is then pushed into the past with an assurance that it will not happen again or has been addressed to the best of the defendant's ability.

This, at least partly, explains the low fines imposed upon defendants; the event is decontextualized, individualized, and sanitized, and then the defendant's good record, cooperativeness, remorse, and subsequent improved work health and safety performance is introduced to the court so that, in the eyes of the court, the defendant is significantly exculpated.

The process outlined above defuses work health and safety as an issue. The court is seen to be dealing with the issue and convicting offenders; but at the same time, sanitizing the underlying issues so that work-related illness and injury are not seen as a consequence of an unsystematic approach to managing work health and safety or the result of unequal power structures, the underlying activity (the production of goods and services) is not threatened. In other words, the court plays a major legitimating role in work health and safety, and ensuring that the underlying issues are largely untouched. Here we can agree with the Robens Report (1972: para. 261), which correctly noted that "the criminal courts are inevitably concerned with *events* (my emphasis) that have happened rather than with curing the underlying weaknesses that caused them," although the argument in this chapter is not just that a prosecution does not cure the contravention that led to the prosecution, but also that a prosecution does not scrutinize social and economic processes underpinning work health and safety.

Of course, to maintain legitimacy, the law must appear to be just and effective; and this is achieved by the mere fact of some prosecutions taking place and the apparently high penalties that are imposed upon some offenders. In the past 15 years, the Australian courts have imposed significant penalties. For example, in 1996, in *WorkCover Authority of NSW v Thiess Contractors Pty Ltd* (unreported, Industrial Court of New South Wales, Full Court, 19 April 1996), the court imposed a total fine of A$175,000 for offenses arising out of three separate incidents in which two employees were killed and a third severely injured. In 2001, after a highly publicized gas explosion that killed two workers and seriously injured eight others brought the State of Victoria to a standstill, the Victorian Supreme Court in *DPP v ESSO Australia Pty Ltd* (2001) 107 IR 285 imposed a total of A$2 million for 11 contraventions of the Occupational Health and Safety Act 1985 (Vic), including a number of maximum penalties to some of the charges. In August 2006, the Victorian County Court fined Foster's Australia Limited A$1.25 million for two contraventions of the employer's general duty in the Victorian Occupational Health and Safety Act 2004. One of the aggravating factors in this case was that there had been a similar incident on a similar machine just over 3 years before the incident giving rise to the prosecution. While there had been improvements to that machine, the machine that had killed the worker in this case had not been altered. These types of penalties are reported in the media and give the impression that prosecution is used to punish egregious offenders.

The process of decontextualizing and individualizing prosecutions is embedded in the form of the criminal law, which has been used without modification for work health and safety offenses (for elaboration of this argument, see Johnstone, 2003: 276–288). The criminal law has traditionally focused on events committed by individuals with guilty intention (*mens rea*). In work health and safety prosecutions, the criminal law is applied, without modification to very serious work health and safety offenses, which do not require proof of *mens rea*. These offenses are usually committed by organizations and are essentially concerned with a failure to take a systematic approach to eliminate risk at work as far as is reasonably practicable. In short, there is a mismatch between traditional, mainstream criminal law and the kinds of criminal offenses found in the work health and safety statutes in Australia and in the UK. Unfortunately, this has had the consequence that there is a dominant view, especially among lawyers, that this signals that work health and safety offenses are "not really criminal" because they do not bear the traditional markers of "real crime" (essentially, guilty intent). The better argument is that the work health and safety statutes themselves state offenses to be "criminal," that putting a worker at risk is a very serious (and criminal) matter, and that rather than arguing that work health and safety offenses are not "really criminal" because they don't resemble "real crimes," criminal law, procedure, and sentencing need to be reconstructed to take account of the nature of corporate crimes and corporate criminal offending.

The Robens Report (1972), and subsequent policymakers in Australia, have failed to take up the challenge that the existing form of the criminal law is inappropriate for work health and safety offenses, and to reconstruct the criminal law so that it is effective to deter and punish work health and safety offenders. The only change to criminal procedure has been to override the effect of the rule against duplicity so that a single charge for a contravention of a general duty offense can include particulars of the defective system of work and inadequate instruction and supervision (see, for example, Model Work Health and Safety Bill, Section 233). This model bill (so called because it has been endorsed by the Australian Workplace Relations Ministers Council and is to be adopted by all Australian jurisdictions) only prevents the multiple splintering of an event in a prosecution and does not in any way overcome the consequences of the event focus. The sentencing principles developed by the courts have, to some extent, tried to ensure that the courts address how far the defendant fell short of its duties to introduce systematic work health and safety management, but they do not specifically address the way in which the event focus decontextualizes work health and safety issues or the types of arguments, outlined above, that defendants regularly use to mitigate their liability. Instead, as the remainder of this chapter shows, the Australian debate about work health and safety enforcement has largely been about equipping inspectorates with a wider range of "enforcement tools," and to some extent, in making the consequences of prosecution more serious. There have been no explicit attempts to reverse the process of conventionalizing work health and safety offenses or of reasserting the true criminality of work health and safety offenses, although, as the rest of this chapter suggests, some of the work health and safety reforms of the last 20 years might be argued to have some impact on that issue.

THE AUSTRALIAN WORK HEALTH AND SAFETY
ENFORCEMENT DEBATE

Since the early 1980s, approaches to enforcing work health and safety statutes have been much debated in Australia. Apart from increasing maximum penalties (as discussed in the previous section), the debates have mainly been pragmatic and instrumentalist, and largely focused on developing a wider array of enforcement methods and sanctions to be used against firms and individuals contravening the work health and safety legislation. Influential in the debates has been Ayres and Braithwaite's (1992: ch. 2) work on responsive enforcement, and also the notion of "risk-based" regulation.

The theory of responsive enforcement (see Braithwaite, 2011) has attracted regulators because it purports to overcome the limitations of both the advise and persuade approach and the punishment or deterrence approach (see Braithwaite, 2002: 32; and for a discussion of the perceived strengths and weaknesses of the two approaches, see Gunningham and Johnstone, 1999: 111–114). It does so by advocating a judicious mix of the two approaches in some form of graduated enforcement response using a hierarchy of sanctions or "enforcement pyramid" (see also Gunningham and Johnstone, 1999; Sigler and Murphy, 1988, 1991; Wright, Marsden, and Antonelli, 2004). The hierarchy of sanctions includes informal advisory and persuasive measures at the bottom, administrative sanctions (e.g., improvement and prohibition notices) in the middle, and punitive sanctions (prosecution) at the top. As Braithwaite puts it, regulators "should be responsive to the conduct of those they seek to regulate," or more particularly, "to how effectively . . . corporations are regulating themselves" before "deciding on whether to escalate intervention" (Braithwaite, 2002: 29; also Ayres and Braithwaite, 1992: ch. 2). The theory posits that credible enforcement must include a significant deterrence component, but targeted to offenders and circumstances when advice and persuasion have failed, and when deterrence is likely to be most effective. Crucial to the theory is the paradox that the greater the capacity of the regulator to escalate to the top of the hierarchy of sanctions, and the greater the available sanctions at the top of the pyramid, the more duty holders will participate in cooperative activity at the lower regions of the hierarchy (Ayres and Braithwaite, 1992: 39).

Critics on the left point to the pluralist assumptions underpinning responsive regulation, "including the idea that power in modern social orders is dispersed rather than concentrated, and that a variety of interests can be mobilised to influence the formal political agenda" (Tombs and Whyte, 2007: 153). In short, responsive regulation, including responsive enforcement, is based on a liberal political agenda, underestimates the extent and seriousness of corporate contraventions of work health and safety regulation, and fails to recognize that the concentration of power in corporations considerably shapes regulator responses to corporate crime (Tombs and Whyte, 2007: 153–157). It is inappropriate to begin with the assumption that employers will strive for virtue, given the structural conflict between capital and labor, and capitalist imperatives to place production ahead of work health and safety (see Slapper and Tombs, 1999).

Other commentators argue that the kind of interactive (tit-for-tat) and graduated enforcement strategy envisaged by Ayres and Braithwaite (1992) is often difficult to operationalize (see Gunningham, 2007: 124–128; Johnstone, 2004b: 158–159). For example, for the pyramid to work in the interactive, tit-for-tat sense envisaged by its proponents, the regulator needs to be expert in systematic work health and safety management, able to identify the kind of firm it is dealing with and to understand the context within which it operates; and the firm needs to know how to interpret the regulators' use of regulatory tools, and how to respond to them (Black, 2001: 20). These are not easy tasks. Second, most work health and safety regulators do not have the resources to work their way up and down the pyramid with each duty holder (see further Gunningham and Johnstone, 1999: 123–129), although this is not true in some industries (hazardous facilities and mining) wherein work health and safety inspectors can visit a single facility up to half a dozen times each year. Further, there is a tension between the tit-for-tat approach in responsive enforcement, in which the severity of sanction is at least partially dependent on how the firm responds to the regulator, and traditional values of law enforcement that emphasize that the sanction must be proportionate to the gravity of the offense.

For at least these three reasons, scrutiny of the "compliance and enforcement" policies of each of the Australian regulators, and of the recently released National Compliance and Enforcement Policy, strongly suggests Australian work health and safety regulators, while claiming to have adopted "responsive enforcement" and working with a hierarchy of enforcement sanctions, in fact simply choose what they consider to be the optimal sanction from the hierarchy of sanctions, based on the extent to which the firm's work health and safety compliance falls short of the level required by work health and safety standards; the resulting level of risk to workers and others; and the attitude, level of cooperation, and prior compliance record of the firm. This is not responsive enforcement in the sense envisaged by Ayres and Braithwaite (1992), but rather a one-off proportional response to the perceived level of offending and the track record of the firm.

Another strong strand in Australian regulatory thinking about enforcement emphasizes the importance of risk-based enforcement to ensure that enforcement strategies are "effective" (ensuring regulatory resources have the maximum effect on outcomes) and "efficient" (reducing administrative and compliance costs, and reducing unnecessary inspections) (Hampton, 2005: 4). A risk-based approach to enforcement means that inspectorates move away from random inspections to more targeted intervention, focusing on firms (and industries) creating the most health and safety risks at work (see Gunningham, 2007. For an overview of different approaches to targeting, see Fooks, Bergman, and Rigby, 2007: 37–41). The Australian Productivity Commission (2010: 118) has concluded that Australian work health and safety regulators take a targeted or risk-based approach to enforcement.

To a large extent, risk-based enforcement and the Australian work health and safety regulators' version of responsive enforcement, described above, dovetail. One of the things they have in common is that they are built upon "consensus" theories of regulation, occasionally verging into a neo-liberal theory of regulation (to adopt the very helpful framework articulated by Tombs and Whyte, 2007: 153–160). Given the starting point for this chapter—the deep entrenchment of the ambiguity and conventionalization of work health and safety crime—it is clear that public discourse about work health and safety enforcement is firmly anchored within a consensus model of regulation, wherein critical perspectives that do not fit the assumptions of the consensus framework rarely find their way onto the regulatory agenda.

Expanding the Enforcement Tools

Since the late 1980s, Australian work health and safety regulators have sought ways of bolstering the sanctions in the work health and safety statutes. As the first part of this chapter has argued, while some contributors to this debate (particularly some unions and academics) have been critical of the historical advise and persuade model of enforcement, the debate has largely been framed by an acceptance of the consensus model of regulation, which is uniformly accepted by employers and work health and safety regulators. For regulators, the issue is largely to develop a flexible array of sanctions that work, in the sense of improving compliance. As a result, an innovative set of sanctions has been introduced, largely to strengthen the middle regions of the hierarchy of enforcement sanctions available to regulators. These new sanctions include the use of infringement notices by regulators and the possibility of accepting enforceable undertakings. Little has been done to address the perception that work health and safety offenses are not really criminal.

The major initiatives with prosecution have been significantly to increase the level of maximum penalties and to introduce a number of important nonfinancial sanctions (such as court-ordered publicity, modified community service orders, and corporate probation). Some governments have flirted with reforms to corporate manslaughter, but only the Australian Capital Territory has implemented these. Most of the other work health and safety statutes have introduced new offenses with higher fines for offenses exposing workers to the risk of serious illness, injury or death, and/or when the offenses involved gross negligence or recklessness. In many senses, the most significant development has been the recasting of the provisions creating offenses for corporate officers. The question, of course, is to what extent do these developments reverse the public perception that work health and safety offenses are not really criminal? As the rest of this chapter argues, the answer to this question is, not much, if at all. Many, it can be argued, further conventionalize work health and safety crime and/or lead to work health and safety offenses assuming an even more ambiguous nature.

Infringement Notices

Administrative penalties applied directly by the regulator are a feature of most work health and safety regulatory regimes around the world (Fooks et al., 2007: 42). From the early 1990s, penalty notices or infringement notices (on-the-spot fines) were introduced in New South Wales, the Northern Territory, Queensland, Tasmania, South Australia, and the ACT for at least some contraventions of the work health and safety statutes. Infringement notices enable enforcement of lesser offenses in a quick, easy, and inexpensive process that bypasses the courts (see Australian Law Reform Commission, 2002: 418; Bluff and Johnstone, 2003). Australian infringement notices tend to impose small fixed penalties: for example, the highest penalty in New South Wales and Queensland is A$1,500; in South Australia, A$315; and in the Northern Territory, A$250. They are also quite inconsistent between jurisdictions in terms of the size of the penalties, the offenses for which notices might be issued,

126 / SAFETY OR PROFIT?

and the persons to whom notices may be issued (see National Review into Model Occupational Health and Safety Laws, 2009: 321–322).

Infringement notices are "intended to be a penalty for a particular episode of non-compliance and highlight that the breach is serious enough to warrant a fine while avoiding court action" (National Review into Model Occupational Health and Safety Laws, 2009: 321). There appears to be some evidence that infringement notices can provide a spur to enforcement action. Gunningham, Sinclair, and Burritt (1998) reported that infringement notices were perceived as an effective means of "getting the safety message across"; that when issued, they were treated as a significant "blot on the record," which spurred preventive activities; and that in some large companies, infringement notices issued were seen as an indicator of the work health and safety performance of managers.

In fact, as Table 6.2 suggests, infringement notices are sparsely used in Australia. They are very much in the consensus, compliance-promoting tradition of work health and safety enforcement, hence their popularity with regulators. The biggest problem with their use is that they tend further to undermine the criminality and promote the conventionalization of work health and safety offenses, because offenses can be addressed by issuing infringement notices that, by definition, are used with lesser offenses (as one inspector observed in a recent study, "Inspectors look like traffic wardens"), and that avoid contact with the principal institutions of the criminal law, the criminal courts.

Enforceable Undertakings

Enforceable undertakings are an Australian invention (see Parker, 2004). They are promises enforceable in court and are "offered" by an individual or firm allegedly in breach of the law. They operate as an agreement between the firm or individual and the regulator, in which the firm or individual undertakes to do or refrain from doing certain activities. The agreement effectively serves as a substitution for, or augmentation of, alternative regulatory enforcement (civil, administrative, or even criminal action). If contravened, the undertaking is enforceable in court. From the early 2000s, enforceable undertakings were made available under the work health and safety statutes in the Commonwealth, Victoria, Queensland, Tasmania, and the ACT, and have been included in the new Model Work Health and Safety Bill 2009, which is to be adopted by all Australian jurisdictions to replace their current work health and safety statutes.

At their most ambitious, enforceable undertakings aim to "build compliance" within a firm by stimulating management commitment, inducing firms to "learn how to comply" and to institutionalize compliance, and by inculcating long-term change within the organizational culture itself (see Johnstone and King, 2008; Parker, 2002: 43–61). They are a potentially important element in responsive enforcement because they can be tailored to the circumstances of the firm and can achieve outcomes that cannot generally be achieved in court, because of the limited range of sanctions traditionally found in the work health and safety statutes and because prosecutions are time-consuming, expensive, and adversarial. They are also said to implement the principles of restorative justice, which emphasizes a collaborative approach in which those with a stake in the offense come together to resolve a negotiated response (see Parker, 2004).

In fact, enforceable undertakings have been little used, apart from in Queensland, where 65 were accepted in the period 2003–2009 (see Johnstone and King, 2008; Johnstone and Parker, 2010). Terms commonly found in Queensland undertakings that were accepted included

- A work health and safety management system implemented and audited to an acceptable standard;
- Other benefits to the workplace and workers (e.g., improved work health and safety training, changed operational practice, and appointment of work health and safety staff);
- Benefits to industry (e.g., development of training or an industry code of practice);
- Benefits to the community (organizing and funding a town health and safety day, funding work health and safety research); and
- Auditing of both the work health and safety management system and the undertaking itself.

Preliminary research (see Johnstone and King, 2008) suggests that enforceable undertakings have the potential to be an effective sanction if properly used. They can have a significant deterrent effect: the monetary value of enforceable undertakings in Queensland in the period 2003–2008 was found to be more than six times the dollar value of the maximum possible penalty for the alleged offenses that gave rise to the undertakings, and the total value of enforceable undertakings was over eight times the total of fines paid in both rejected and withdrawn matters combined (Johnstone and King, 2008). The same research found indications that the effects of the enforceable undertaking process include a more systematic approach to work health and safety management, increased resources allocated to work health and safety, cultural changes within organizations, increased worker participation, direct cost savings and efficiencies, and a reduction in workplace incidents.

There are, however, good reasons to be wary of enforceable undertakings. The first difficulty is that to be used effectively, regulators have to be fully transparent and accountable and must monitor undertakings once they are accepted (see Parker, 2004). While Queensland has made some progress with these issues, generally they are still areas of weakness for Australian work health and safety regulators.

The more significant concern with enforceable undertakings is that they are clearly aligned with a consensus or compliance regulatory framework and exacerbate the ambiguous nature of, and conventionalize, work health and safety offenses, because they enable firms to evade prosecution by offering an undertaking. Regulators have sought to address this by indicating they will not accept undertakings for very serious offences.

128 / SAFETY OR PROFIT?

Steps to Strengthen the Criminality of Work Health and Safety Offenses

The two measures discussed in the previous section, infringement notices and enforceable undertakings, are important because they enable work health and safety regulators to design enforcement regimes so as to avoid a compliance deficit; or as the Macrory Report (2006: 24) put it, "where non-compliance exists and is identified but no enforcement action is taken because the appropriate tool is not available to the regulator."

But as I have already argued, this is a regulatory concern played out within a consensus framework of regulation and seeks to find regulatory tools that work. In the rest of the chapter, I examine aspects of the sanctions debate that have the potential to strengthen the criminality of contraventions of work health and safety statutes.

Appleby (2003) suggests that in England and Wales, it would be better for work health and safety to "rehabilitate the status" of the Health and Safety at Work Act 1974 for many industrial deaths by emphasizing its breach as truly criminal, as work health and safety law is in fact criminal law. I agree with this proposition, except I would argue that it should apply to all kinds of work health and safety general duty offenses placing workers at high levels of risk, even when a fatality does not result.

Higher Penalties

One possible way of increasing the criminality of work health and safety offenses is to signal the seriousness of work health and safety offending by increasing the financial penalties that the courts can impose for proven work health and safety offenses. To a significant extent, Australian governments have sought to ratchet up penalties for breach of existing duties: for example, by the end of 2011, the maximum fines were A\$1,020,780 for corporations in Victoria; A\$550,000 for corporations in New South Wales (\$825,000 for repeat offenses); A\$500,000 in Western Australia (A\$625,000 for repeat offenses); A\$300,000 in South Australia; and A\$550,000 in the Northern Territory. The Model Work Health and Safety Bill appears to have increased maximum fines even more: the maximum fine for a corporation will be \$1.5 million if the contravention exposes an individual to a risk of death or serious injury or illness (when there is no serious risk, the maximum penalty is \$500,000). This is a step in the right direction, although these maximum penalties are still not large enough to strike fear into large corporations. The difficulty is that merely increasing the maximum fines does not address the processes of decontextualization and individualization that take place when offenders are sentenced, and which were outlined earlier in the chapter.

There have been other attempts to reconstruct the criminal law to address the types of "systems-based" offenses committed by corporations under the work health and safety statutes. Recognizing the well-documented weaknesses of the fine—principally that the cost of fines can be passed on to others, that the fines imposed are not calculated with reference to the means of the defendant, and that fines don't require offenders to directly improve work health and safety—Australian governments have introduced new corporate sanctions, including

- adverse publicity court orders (New South Wales, Victoria, South Australia, the ACT, and the Northern Territory),
- a court order that the offender participate in a work health and safety-related project (New South Wales, Victoria, South Australia),
- an order requiring the defendant to take remedial measures (Commonwealth, New South Wales, the ACT, and the Northern Territory) or to undertake training (South Australia and the Northern Territory),
- an order adjourning the case with or without conviction and requiring the defendant to undertake to not reoffend within 2 years and to engage a consultant, develop systematic work health and safety management, and to be monitored by a third party (Victoria and Western Australia).

The Model Work Health and Safety Bill has included versions of each of these nonfinancial sanctions (see Part 13, Division 2). While these are important developments, there is a danger that the new sanctions will not be used much; certainly this has been the experience with the existing nonfinancial sanctions to date. Anecdotal evidence from judges and prosecutors in New South Wales suggests that both judges and prosecutors, unused to these new sanctions, wait for the other to initiate discussion of them in proceedings.

A further problem is that some regulators argue that these nonfinancial sanctions themselves signal that work health and safety offenses are not really criminal because they are not traditional criminal sanctions. Against this, it can be argued that these reforms are part of the process of adjusting the criminal justice system to corporate offending. Further, some of the sanctions (e.g., adverse publicity orders) can inflict significant reputational damage to offenders; and other sanctions, particularly those that require offenders to assure the court that they will introduce structural changes within their organization, implementing a traditional rationale of the criminal law, namely, rehabilitation.

A third concern, and one linked to the second concern, is that both nonfinancial sanctions and increased financial penalties will only strengthen the criminality of work health and safety offenses if imposed *after conviction*. In some Australian jurisdictions, and in the Model Work Health and Safety Bill, courts can impose penalties without conviction. It is difficult to think of a clearer indication that work health and safety offenses are not really criminal than imposing regular penalties (financial or nonfinancial) without convicting the offender of the work health and safety offense than the court had found to be proven. In some Australian jurisdictions, Queensland being the prime example, it is standard practice for fines to be imposed without conviction, apparently with the agreement of the prosecutor.

130 / SAFETY OR PROFIT?

The Manslaughter Debate

The reinvigorated work health and safety debate in Australia has, since the mid-1980s, included regular calls for manslaughter prosecutions for work-related deaths (see Hall and Johnstone, 2005). Unlike the offenses against the general duties in the work health and safety statutes, which are absolute liability (qualified by reasonable practicability) offenses and do not require an injury or death for the offense to be committed, manslaughter prosecutions can be brought only when a fatality has occurred and when the person causing the death was guilty of criminal fault. In Australia, at common law, the requisite fault is "criminal negligence"; that is, "a great falling short of the standard of care" that a reasonable person would have exercised, involving "such high risk that death or grievous bodily harm would follow that the doing of the act merited criminal punishment" (see *Nydam v R* [1977] VR 430 at 445). Manslaughter prosecutions can be investigated only by the police and prosecuted by the relevant Director of Public Prosecutions.

There are technical legal reasons why it is difficult to prosecute corporations and corporate officers for manslaughter, and consequently, manslaughter prosecutions for workplace deaths are extremely rare. Much of the debate is about changing the legal rules to ensure that manslaughter can be more easily attributed to corporations and corporate officers (see Johnstone, 2004a: 464–472). Essentially, there are three issues in the manslaughter debate: should the grossly negligent manslaughter elements be clarified or replaced by something else?; should the rules be changed to enable senior corporate officers to be prosecuted?; and should the rules attributing liability to corporations be reformed?

Some have argued that there is little evidence that manslaughter prosecutions will have a deterrent effect and that manslaughter prosecutions should be used for symbolic purposes (see Haines and Hall, 2004). Others have argued that manslaughter prosecutions may be counterproductive, in that by focusing on a few particularly "serious cases" and singling them out for special treatment, the regulator risks further undermining the criminality of work health and safety offenses (see Carson and Johnstone, 1990: 140). "By prising out a few cases for treatment under separate, criminal auspices, the criminal status of what is left is rendered even more ambiguous than it is already becoming under the impact of the continuing historical and structural processes" (Carson and Johnstone, 1990: 140), outlined earlier in this chapter. While industrial manslaughter provisions were debated in the 1990s, the only manslaughter provisions introduced in Australia were in the Australian Capital Territory in 2003 (the Crimes [Industrial Manslaughter] Amendment Act 2003 [ACT]).

A more common response to the debate has been for governments to introduce provisions imposing higher maximum penalties for contraventions of the general duty provisions in the work health and safety statutes when an incident resulted in death, serious injury, or a serious risk of death or serious injury, and/or when there was an element of *mens rea*. To give two examples, Section 32A of the New South Wales Occupational Health and Safety Act 2000 created an offense when a breach of a general duty provision caused the death of a person to whom a duty is owed and the duty holder was reckless as to danger of death or serious injury. The maximum penalty for this offense was A$1.65 million for a corporation and A$165,000 or five years imprisonment for an individual person. Section 32 of the Victorian Occupational Health and Safety Act 2004 creates an offense of recklessly placing another person at a workplace in danger of serious harm, with a maximum penalty of A$1,020,780 for a corporation and A$204,156 or five years imprisonment for an individual person.

It could be argued that this is a better approach to reasserting the criminality of work health and safety offenses: by having penalties that escalate as the level of risk from the offense increases, and penalties that escalate with the *mens rea* of the offender. In other words, rather than perpetuating the distinction between work health and safety offenses as quasi-criminal or regulatory crime and "mainstream" crimes as being real crime, this approach emphasizes the criminality of the work health and safety offenses. A problem, of course, is that this approach tends to emphasize criminality when the offenses assume characteristics well known to the traditional criminal law—a serious injury or a death and *mens rea*—rather than reconstructing it. A better approach might be to enact the health and safety offenses in the mainstream Criminal Code (see Carson and Johnstone 1990).

Liability of Officers

A corporation is an artificial entity and can operate only through its human agents, including its officers and workers. The Australian work health and safety statutes have, since the 1980s, included provisions enabling senior corporate officers to be prosecuted when the corporation in which they are an officer commits a work health and safety offense. Until recently, the models used have been based on derivative liability: that is, the issue of officer liability arises only once a corporation has committed an offense.

One model was "accessorial" liability, the approach taken in Section 55 of the Occupational Safety and Health Act 1984 (WA), which provides that a director, manager, secretary, or other officer of the body corporate is also guilty of an offense committed by the body corporate if the offense occurred with the consent or connivance of, or was attributable to any neglect on the part of, the officer. A similar approach is taken in Section 37 of the Health and Safety etc At Work Act 1974 (UK). Accessorial liability of this type is difficult to prove.

A second approach was the "imputed liability" approach taken in Section 26 of the Occupational Health and Safety Act 2000 (NSW), Section 167 of the Workplace Health and Safety Act 1995 (Qld), and Section 53 of the Workplace Health and Safety Act 1995 (Tas). Here, officers were deemed to have committed offenses committed by the body corporate unless the officer satisfied the court that he or she was not in a position to influence the conduct of the corporation in relation to its contravention of the provision; or he or she, being in such a position, used all due diligence to prevent the contravention by the corporation.

132 / SAFETY OR PROFIT?

A third approach (for example, Section 144 of the Occupational Health and Safety Act 2004 [Vic]) is also built on imputed liability, but couched the director's duty in terms of "reasonable care." A similar approach is taken in South Australia.

Only Queensland and the Northern Territory provides for custodial sentences. The possibility of imprisonment certainly helps assert the criminality of the work health and safety statutes, but very low prosecution rates conventionalized these offenses.

A new approach has been taken in the Model Work Health and Safety Bill. Instead of relying on imputed or accessorial liability, the model bill imposes a positive and proactive duty on "an officer" or "a person conducting the business or undertaking" to "exercise due diligence to ensure that the person conducting the business or undertaking complies with that duty or obligation." Due diligence is codified in the model bill. Why this recasting of the duty is so important is that every officer must exercise due diligence, and the officer can be prosecuted if he or she does exercise due diligence, even if the work health and safety performance of the firm is satisfactory. The maximum penalties for a contravention of the officer's duty are significant: A$100,000 for ordinary contraventions; A$300,000 if the failure to exercise due diligence exposes an individual to a risk of death or serious injury or illness; and A$600,000 and the possibility of 5 years imprisonment if the officer is reckless.

Enforcement Powers for Workers

Finally, Nichols and Armstrong (Nichols, 1997, ch. 3) reminded us in their pamphlet of the need to strengthen the power of workers to resist production pressure and to work safely. The Australian work health and safety statutes have made some headway in arming workers and their representatives with enforcement powers, though the approach has not been consistent.

Before the introduction of the Model Work Health and Safety Bill, five of the nine work health and safety statutes empowered health and safety representatives (HSRs) to issue provisional improvement notices; and, as part of a broader issue resolution process, four gave HSRs power to direct that dangerous work cease.

A different approach was followed in New South Wales, which, at least since 1983, has given authorized union officials the right to investigate suspected offenses against the work health and safety statute and has enabled trade union secretaries to bring work health and safety prosecution. About 20 prosecutions have been brought since 1983, many of which have been groundbreaking. One concerned inadequate staffing during a strike at a government school for disabled children resulting in an assault on a teacher; and another three addressed the trauma suffered by Finance Sector Union members during armed bank robberies and led to the introduction of barriers, resulting in a significant decline in both robberies and affected staff.

The Model Work Health and Safety Bill has vested all HSRs with the right to issue provisional improvement notices and to direct that dangerous work cease as part of an issue-resolution process. It also contains provisions for trade union officers or employees with work health and safety entry permits to investigate suspected contraventions of the model bill or to consult and advise "relevant workers" (i.e., workers who are members, or eligible to be members, of the union and who work at the workplace) on health and safety issues. If contraventions are found, the union entrant has no right to enforce (including prosecution) and only has the right to warn persons at risk of the health and safety risk. The New South Wales version of the model bill has, in fact, retained a restricted right of union prosecution.

These provisions are important because they vest enforcement powers in the workers who bear the work health and safety risks of work. They also increase the accountability of regulators and of employers. Only in New South Wales is there a decent hierarchy of sanctions available to workers and their representatives to aid enforcement. While these sanctions do not necessarily address the issue of the criminality of work health and safety offending, they do arm workers with some measures to resist production pressures from overwhelming health and safety concerns.

CONCLUSION

This chapter has argued that despite major debate in Australia about improving enforcement of the work health and safety statutes, the decriminalization of health and safety offenses has barely been addressed. Part of the difficulty is that the quasi-criminal status of work health and safety offenses is deeply entrenched in public discourse about health and safety regulation; even progressive commentators tend to argue that health and safety offenders should be criminalized by launching more manslaughter prosecutions (see, for example, Glasbeek, 1998; Slapper, 2000; Toombs and Whyte, 2007), rather than by reconstructing the criminal law surrounding health and safety offenses. The other difficulty is that the discourse surrounding health and safety enforcement, particularly by regulators, is firmly anchored in a consensus regulatory framework, in which the emphasis is on "promoting and securing compliance" rather than on deterrence and punishment. Just how strong this approach is among regulators is illustrated by the Model Work Health and Safety Bill, which frequently uses the term *compliance* for *enforcement*. The central concern of regulators is to find an armory of enforcement tools that work to improve compliance—the issue of reasserting the criminality of health and safety offenses is off the agenda.

PART III

The Role and Limits of Evidence

Today, the call for evidence-based policy is a familiar one. Who, after all, would publicly endorse policy that was not evidence based? Despite the rhetoric however, in the field of health and safety at work, policy is often formulated on the basis of very poor evidence. The Robens Report (1972) relied on hardly any evidence as ordinarily understood (as opposed to business opinion), and the situation is not much better these several decades later. In Chapter 7, Andrew Watterson examines the case of occupational cancer prevention in the UK and demonstrates how diverse interests impede progress, something that is not helped, either, by the general neglect of worker testimony. He provides a case study of the particular relation between night work and cancer, and he urges the general case for the adoption of the precautionary principle when inspecting new data and that the public health response should be one that fails to safety, not danger.

The failure to give sufficient weight to worker opinion is also a feature of Chapter 8, in which Wayne Lewchuk examines the workings of a review panel on the effectiveness of health and safety regulation in Ontario, Canada, conducted in the context of the spread of precarious work. The argument advanced in the report that any further regulations must be cost-effective and that evidence-based changes must be made so as to also strengthen competitiveness is entirely familiar, as is opposition to a return to a regulatory regime based on external enforcement. The neglect of worker voice is again underlined by Charles Woolfson in his account of the 2010 *Deepwater Horizon* disaster in the Gulf of Mexico. More generally, the chapter provides an example of the extreme application of self-regulation to the extent that industry, in this case offshore oil, succeeded in its mission of regulatory capture, resulting in the maximum scope for business discretion and a minimum of external accountability. As Woolfson argues, the regulatory arrangements in force in the U.S. offshore industry essentially replicated those that had existed before the 1998 *Piper Alpha* disaster in the North Sea, which were later modified following the 1990 review of the situation in the 1990 Cullen Report. The failure to learn from the evidence of *Piper Alpha*, like much else, cannot be understood in the absence of taking into account the conflict between safety and profit or, in this particular case, the combination in one agency of responsibility for the regulatory oversight of the industry and the collection of revenues from it.

http://dx.doi.org/10.2190/SOPC7

CHAPTER 7

Competing Interests at Play?
The Struggle for Occupational Cancer
Prevention in the UK

Andrew Watterson

INTRODUCTION

> Occupational cancers represent a challenge to industry as well as public health
> agencies, as they are the only cancers the development and occurrence of which
> can be largely or completely eliminated, if proper precautionary measures are
> taken to prevent any undue contact by the workers or if the cancerigenic (now
> called carcinogenic) factors are excluded from industrial operations (Hueper,
> 1942: 9).

Occupational diseases, including cancers, remain a relatively neglected research
and policy priority for researchers from the sciences and social sciences despite
their public health importance. This chapter describes the toll taken by occupational
cancer primarily in the UK and examines how different types of evidence are and
have been weighted in the contested policy process to address that toll. Recognition
of occupationally caused and related cancers and estimating occupational cancer
burdens has proved difficult in the UK but should not have constrained some
prevention interventions. The terrain is highly contested, not simply in terms of
numbers but also in terms of diagnosis and state listing of the diseases. The role of
groups such as employers, scientists, politicians, regulators, trade unions, and
workers in the process is explored. Worker testimony on and concern about occupational cancers has often been ignored but form an important part of some initiatives
to prevent such diseases. A case study on occupational cancer and nightshift work
is used, drawing on UK and European research and policy responses, to illustrate
the complex competing interests and highlight the different responses within and
beyond the UK to occupational cancer recognition and prevention. This chapter
examines why this is and what is being done to address the neglect. Using research

138 / SAFETY OR PROFIT?

from World Health Organization (WHO) bodies and academic institutions, trade unions, and civil society bodies at an international and UK level have begun to work out preventative strategies to address the social, human, and economic costs of occupational cancers.

Occupational physicians have for centuries described occupational cancers and some, like Hueper (1942), addressed the problems of prevention directly. In *Safety or Profit*, Theo Nichols and Peter Armstrong (1973) noted that most industrial sociologists simply ignored job-induced illness as well as injuries. Workers have always, if not widely, chronicled and discussed the impact of the workplace on their lives, but Nichols' studies of chemical workers began to fill an important gap and reveal a complex picture of pressures, problems, and locally mediated solutions to owners and management challenges on working conditions, including health and safety (Nichols and Armstrong, 1976; Nichols and Beynon, 1977).

That work helped to develop far more rigorous social science and informed related activity on health and safety at work, but it is still all too often a marginal research activity uncoupled from wider debates about poverty and inequality. There were exceptions. In 1979, for example, Peter Townsend included physical working conditions as an indicator of deprivation that hit certain populations and groups but noted the lack of interest in the topic among many trade unions and their focus instead on pay, employment, and working practices (Townsend, 1979: 43). Such an approach still remained relevant later in the 20th century. The 1998 CAREX report for Great Britain, for example, concluded that workplace exposures to carcinogens were restricted to about one fifth of the working population, mainly unskilled, semiskilled, and skilled workers (Kauppinen, Toikkanen, Pedersen, Young, Ahrens, Boffetta, et al., 2000). Other nonmedical commentators identified particular carcinogens as threats to health: for instance, Linhart, in his descriptions of the carcinogen benzene in France's car plant at Choisy during 1968 (Linhart, 1981: 40, 44–45), where "all the workforce knows that the Citroen doctors get higher bonuses in proportion to the small number of people they put off work."

In 1978, Samuel Epstein, a pathologist in the United States, traced the growing threat from workplace and wider environmental carcinogens. He dissected the neglected political as well as scientific competing interests that shaped industrial and governmental responses to occupational cancer statistics linked to calls for controls and prevention (Epstein, 1979). Studies of the political economy of health in the UK in the 1980s drew directly on the Epstein analysis and recognized the competing interests at play, sometimes specifically in the occupational cancer field (Doyal, 1979; Doyal et al., 1983). Unlike the United States at the time, the UK regulation of workplace carcinogens was "generally fragmented and conducted in private" and operated "on the basis of consensus between government-appointed experts rather than the more adversarial, open and participative approach which has developed in the USA" (Doyal et al., 1983: 49–50).

Little has changed. Although at one stage, UK occupational health and safety looked likely to be reprioritized under a broader public health umbrella, the opposite result emerged with a downplaying of the work agenda. Instead, health promotion in the workplace was linked to lifestyle and victim-blaming strategies and ate up scarce resources that had previously been available for mainstream occupational health and safety prevention work (Watterson, Gorman, and O'Neill, 2008). For many health promoters, cancers are perceived as due almost solely to diet, alcohol, smoking, and a lack of exercise, with lifestyle choices creating the major problems and not those industries producing, selling, and using carcinogens that affect employees and consumers alike. There are therefore strong ideological influences producing assessments that the work and wider environmentally related cancer problem in the UK, and indeed globally, is of little significance. This is against a background of considerable cancer morbidity and mortality, and for some cancers, both morbidity and/or mortality statistics have grown throughout the 20th century, for example breast cancer in women and prostate cancer in men (Clapp, Howe, and Jacobs, 2006). When cancers are very unusual and specifically related to occupations or exposures to specific substances—such as the cancer mesothelioma due to asbestos exposure or the liver cancer angiosarcoma due to exposure to vinyl chloride monomer—then recognition, diagnosis, and recording may be relatively rapid although far from perfect (Carlin, Knight, Pickvance, and Watterson, 2008). Lung cancers due to asbestos exposure have taken longer to gain industrial disease recognition in the UK and will be missed more often because of confusion with other causes of this cancer, and epidemiologists now accept that there is at least one lung cancer case caused by asbestos for every mesothelioma case.

The most vulnerable workers also frequently pass under the cancer health recording radar. This is so in terms of moving from job to job, doing precarious and hazardous work, migrating and working in factories and occupations that are not registered or inspected by either the Health or Safety Executive (HSE) or other regulators in local authorities. When they do become visible, the recognition of many occupational cancers by governmental, insurance and legal schemes has often been resisted for decades, and compensation has frequently been minimal (Dalton, 1979; O'Neill, Pickvance, and Watterson, 2007). Cancers sometimes, but not always, may take a considerable time to develop and so the linkage of cause and effect is far more problematic than with injuries at work. However, there are major discrepancies between countries in terms of the latency periods and exposure times recognized for particular occupational cancers. For example, Germany recognized benzene-related leukaemia after 6 months exposure and 2 years latency; and wood-dust-related nasal adenocarcinomas after 5 years exposure and 8 years latency (Popp, Bruening, and Straif, 2002). The UK schemes require much longer exposure to benzene, and the HSE believed in the mid-2000s that wood dust exposure latency for some cancers was 20 years (O'Neill et al., 2007: 437). Additionally, several shipbuilding and engineering firms, insurers and their lawyers have contested compensation cases for asbestos-related cancers knowing full well that victims will die before cases are resolved (Dalton, 1979: 232).

The UK government has run down its occupational health staff and services over several decades and adopted the so-called better regulation agenda. In practice,

140 / SAFETY OR PROFIT?

this has meant softer regulation and on occasion, no effective regulation at all, linked with minimal or no enforcement backed up by low or no fines and nonimprisonment of employers responsible for exposing workers to carcinogens and causing occupational cancers (O'Neill et al., 2007). There are alternative approaches that have sometimes succeeded, connected to strong enforcement and based on both scientific evidence and recognition of social and environmental justice priorities. For example, Italy recently sentenced company directors to 16 years in prison for killing workers in their employment who contracted occupational cancers (Euronews, 2012).

Historically, bodies like Science for People and particular trade unions identified and campaigned for action on UK occupational cancer, but worker testimony on and concerns about occupational cancers were often dismissed by employers and governmental scientists as anecdotal and were thus ignored. Vested interests dominated for a variety of reasons linked to the power of those industries adopting a neo-liberal stance (Michaels, 2008). This has been compounded by the soft regulation and/or no enforcement position on many occupational cancers (O'Neill et al., 2007). Such responses ensure that narrow capitalist interests and values rather than broader civil society ones win out.

The global picture on occupational diseases remains serious, with occupational cancers a major cause of workplace mortality, with almost double the mortality of fatal injuries and violence, as the Chart 7.1 illustrates.

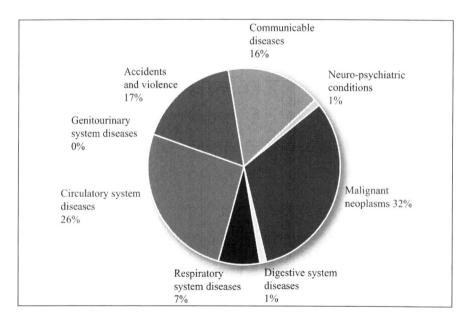

Chart 7.1 Global estimated work-related fatality by cause.
Source: Hämäläinen, Takala, and Saarela, 2007.
The material reproduced with permission of John Wiley & Sons, Inc.

The current peculiar position of policymakers and regulators who produce reports on the economic costs of occupational diseases in the UK is effectively to ignore these reports and maintain inaction. Health policy in the UK is put forward as evidence-based, but the resources, staffing, and enforcement action devoted to occupational cancer prevention do not match the international estimates of the extent of the problem. When scientific evidence establishes both hazards and risks, and when such evidence indicates the importance of adopting precautionary policies, then preventative action on carcinogens in the workplace should be straightforward, whether they be they chemical, biological, physical, or even work organization-related. This is not the case. The reasons why relate to competing and dominant interests in UK society.

The Occupational Cancer Problem

The fractions of cancers due to work and wider environments remain much disputed. There has often been far more debate about the exact number of cancers than the strategies needed to reduce or remove exposures to known carcinogens. This has been good for scientists researching the problem but bad for those exposed in the workplace. Related cancer morbidity tends to be ignored because of the difficulties of reporting. In the 1980s, Doll and Peto's best estimate for occupational cancer deaths was 4% and for pollution and geographical factors, another 5%—a 9% total in the UK (Doll and Peto, 1981). They called for action, though none of any significance occurred. Yet the Doll and Peto figures are gross underestimates and were challenged at the time (Clapp et al., 2006; Fritschi and Driscoll, 2006; Hämäläinen et al., 2007; O'Neill et al., 2007). The HSE's latest estimate of occupational cancers runs at 4.9% and acknowledges that the figure remains an underestimate (Rushton, Bagga, Bevan, Brown, Cherrie, Holmes, et al., 2010).

Even using the 4% outdated figures, this is a significant yet often neglected public health burden, especially when each UK occupational cancer case costs an estimated £2.46 million (O'Neill et al., 2007). Using the 1981 figure, for example, produces more occupational cancer deaths in Scotland each year than murders and road traffic fatalities combined. A 12% figure would mean more such deaths than murders, road traffic fatalities, and suicides combined (Watterson et al., 2008). European Agency staff, as Chart 7.2 shows, estimate 13.6% for males and 2.1% for females (Takala, 2008). The International Agency for Research on Cancer (IARC) estimates 7%–19% of cancers worldwide are due to toxic environmental exposures (Straif, 2008; WHO, 2009). The estimated size of the UK occupational cancer problem is shown in the Table 7.1.

The World Health Organization has identified primary prevention as the key approach to reducing occupational and environmental cancer and promulgated this at several events (Kim et al., 2008; Landrigan, Espina, and Neira, 2011; WHO, 2009, 2011). The response to this approach from individual countries has been slow and frequently hostile.

142 / SAFETY OR PROFIT?

Region	Number of all cancer deaths		Attributable fraction relation to work		Number of deaths attributable to work		Work-related cancer
Cancer (total)	Men	Women	Men	Women	Men	Women	Total
EU 15	528,953	410,829	13.8	2.2	72,996	9,038	82,034
EU 25	600,508	464,757	13.7	2.2	82,194	10,144	92,338
EU 27	623,709	481,307	13.6	2.1	85,106	10,177	95,581
World	3,872,766	3,062,008	9.6%				665,738
Source: Takala, 2008.							

Chart 7.2 Number of cancer deaths, and number and fraction attributed to work by region and gender.

Table 7.1 Estimates of the Proportion of Cancer Attributable to Occupation

Occupational cancer cases/year (GB)	HSE/Doll-Peto estimate	Corrected estimate (Clapp et al., 2006)
% of all cancers	4% (range 2%–8%)	12% (range 8–16%)
Deaths (Lower/upper estimates)	6,000 (3,000–12,000)	18,000 (12,000–24,000)
New cases (Lower/upper estimates)	10,800 (5,400–21,600)	32,000 (21,600–43,200)

Source: Hazards, 2009.

A TOPOGRAPHY OF COMPETING INTERESTS

The topography provided here and in Table 7.2 may oversimplify what can at times be complex and contradictory interactions. However, it does describe the spectrum of competing interests that emerge in the world of occupational health and safety and should inform any analysis of occupational cancers and related cancer prevention efforts in the UK.

Companies and shareholders want profits that may of course be mediated by concerns about reputation and liability. Governments, when captured by neo-liberal agendas, want to support companies, free markets, and deregulate controls. They may do this crudely or subtly. Civil servants and local government staff, be they scientific advisors or regulators, are paid to carry out the ideological wishes of their masters. Few dissent because dissent would often mean dismissal. This has happened in the past to civil service whistle-blowers and scientific expert group members. Governments and their agencies fund the occupational health and safety research that addresses the questions they want answered and in the way they want them answered (Watterson, 1999a). This may relate to soft regulation, to quiescence when, for example, the HSE mission was altered to include economic and business interests as well as worker health and safety. Rarely do HSE professional values and ethics appear to prevent the introduction of weaker workplace health and safety controls. Researchers want research funds and status, although funders may sometimes want token service-user input. Health and safety professional bodies have far more autonomy and may speak out against threats to health and safety, although rarely do they strongly pursue either occupational illness issues in general or occupational cancer cases in particular.

Charities seek funds to support their research and policy work. Just like businesses, some charities assess success based on the size of their income streams. Some cancer charities also spend most of their funds on lab-based pharmacological research geared heavily to treatment and very little on prevention (Alliance for Cancer Prevention, 2012).

Trade unions want to recruit and retain members, negotiate good wages and job security for them, and sometimes too they reflect neo-liberal and corporate values. They also want to reflect their members' workplace concerns and priorities. At times, they will therefore pick up and pursue workplace health and safety; at other times, they may neglect those concerns and prioritize jobs and pay (Watterson, 1999b). Trade union members want good work that includes good health and safety; something that remains a major priority for them in all recent surveys (see Table 7.2).

RESPONSES TO CALLS FOR OCCUPATIONAL CANCER PREVENTION: MANAGING THE PROBLEM AWAY

It has not been in the interests of politicians, employers, and even some civil society groups like cancer charities to research, recognize, and compensate workers for occupationally caused and related cancers and even less to develop effective cancer prevention programs in the workplace. Employers and their insurance companies have externalized costs of occupational cancers; and some, but not all, governments within the UK have protected many employers who create hazardous work. The human, social, and economic costs of occupational cancers are therefore picked up by society as a whole and more particularly by the NHS, the victims, their families, and communities (Watterson, Gorman, Malcolm, Robinson, and Beck, 2006). There are even cases in which chemical and pharmaceutical companies produce or use carcinogens and so expose workers; then they sell carcinogens to the public, and finally, they produce cancer-treatment drugs (Armstrong, Dauncey, and Wordworth, 2007; Epstein, 1978: 79, 150, 506). This cycle of production and profit is frequently condoned by government, whose interests can be close to such manufacturers. Science has often been manipulated, funding skewed, decision makers influenced, and critics attacked by industries to allow companies to continue

144 / SAFETY OR PROFIT?

Table 7.2 Competing Interests that Affect Cancer
Prevention Activities

Sector	Activity	Competing interest
Capital—varies from sector to sector and company to company Pharmaceutical subinterests	Profit	Regulators and civil society. Sometimes trade unions, civil society, and media
Managers	Control of workplace for profit or "efficiency" and ease	Employees, trade unions, public, users
Government	Sometimes supports capital and economic interests	Civil society, voters, opposing voices supporting regulation, enforcement, and high standards
Regulators, including the scientists who conduct research for government	Carry out government policy, including cuts in staff function, and enforcement	Workers, possibly employers and civil society
Trade unions—vary as do workplace responses	Recruit and represent members. Defend jobs	Sometimes employers, governments, members, and civil society bodies
Civil society organizations Environmental groups	Public health and worker health	Capital, sometimes government, sometimes regulators
Professional bodies of health professionals and scientists who may conduct research independent of government and industry or may not	Produce policies and research reports	Other professional bodies, the public as users, civil society
Cancer charities	Income generation	Other charities, civil society bodies
Media	Inform and shape opinions of readers, viewers, and listeners, and respond to needs and requirements of advertisers and owners	Possibly advertisers, owners, and governments

to make, use, and sell carcinogens when the science indicates they should not. Dr. David Michaels, the current head of the Occupational Safety and Health Administration in the United States, has described this process in terms of the "Enronization" of science by industries. Such industries create doubt about the science that identifies their products as risks and hazards. Hence, industries produce doubt as a product to ensure their carcinogens can continue to be used (Michaels, 2008).

In the UK, industry, and often regulators, have deployed a range of arguments to maintain inaction on workplace carcinogens (O'Neill et al., 2007). Some suggest that occupational cancers are a disease of the past, thus downplaying or ignoring current threats from, for instance, diesel, silica, asbestos, and possible nanocarcinogens. They indicate that controls and enforcement are much better now, but when a report appeared on a major established carcinogen such as MbOCA, a bladder carcinogen, the HSE found controls and personal protective equipment (PPE) were inadequate, training was poor, and exposure levels were unacceptable (Keen, Coldwell, McNally, Baldwin, and McAlinden, 2010). They state that cancers are diseases of the older person and not work-related, yet with longer working lives and greater numbers of jobs undertaken, often in a wider variety of workplaces, older workers may be exposed to greater numbers of carcinogens than in the past and may prove more vulnerable. In addition, some cancers in younger people are increasing, and workplace fetal exposures may be part of a growing toxic soup of carcinogens. The final response, which might be termed *paralysis by analysis*, is often "we are reviewing the position, wait and see." This is against a background of Doll and Peto (1981) indicating action was needed in 1980, several books and trade union reports in the early 1980s calling for action on occupational carcinogens, and the running down of the UK government's occupational health services from a substantial and active group in the 1970s to a very small group of staff now (Greenberg, 2005). It is easy to identify which competing interests dominated.

The role of civil society groups in the occupational cancer debate can be highly contradictory. Cancer charities generate income on the basis of funding research primarily, but not exclusively, on treatment; some cancer charities and patient's groups may receive funding from commercial companies (EPHA, 2012; Heinrich, 2012). This mirrors the pharmaceutical industry search for profitable products to sell to the NHS for cancer treatment and is often presented with an exclusively lifestyle message on causation. Little mention of life circumstances and socio-economic inequalities emerge, yet these are the most powerful factors in exposure to and prompt diagnosis of several cancers. Multinational companies such as Ford and Kentucky Fried Chicken have capitalized on opportunities for corporate promotion by supporting fundraising activities like pink ribbon breast cancer events run by breast cancer charities (Heinrich, 2012).

Yet bodies such as the IARC reveal that plastics in vehicles may be carcinogens or have been carcinogenic to their workers; and the food industry may also produce carcinogens that could be removed or reduced in foods. Diesel has been identified as a proven human carcinogen, benzene in gasoline is a carcinogen, frying food in various oils may produce carcinogens, and fatty foods in large quantities that lead to obesity may increase cancer risks (O'Neill et al., 2007). Once more, the commercial interests and marketing drown out the science and community concerns. Some smaller charities and campaign groups do not always go down the treatment route and campaign on governmental and industry removal of carcinogens. They usually have limited resources and minimal direct governmental lobbying power, but they are independent of industry or government and not captured by neoliberal ideologies (Alliance for Cancer Prevention, 2012). Their work has been marginalized but not silenced by the bigger competing interests.

Many workers too will face quadruple cancer risks, but these can be disaggregated and ignored by governments. For example, workers in chemicals and plastics plants may be exposed to carcinogens at work, which may pollute air, water, and soil and affect the workers in their domestic and public environments; they could be present in their leisure pursuits and may be absorbed through food and drink. Official risk assessments, however, may downplay combined effects and low-level exposures.

The role of trade unions, though far from homogenous, has generally been tardy and ambiguous with regard to carcinogens. There was some formal trade union activity by unions like ASTMS and GMB on the topic in the 1980s (ASTMS, 1980; Doyal ct al., 1983) and latterly by GMB and Unite. With declining membership and amalgamations in the following decades, health and safety has not been prioritized, and union health and safety staff and departments have often been cut or marginalized. Few unions have run cancer prevention campaigns in the past, but several now publicize lifestyle approaches to cancer. Nevertheless, some employees not in trade unions, or within trade unions that may be inactive on health and safety, still pursue action on cancer prevention ahead of their leaders (Carlin et al., 2008; Watterson, 1993, 1999b). Some trade unions, placing wages and conditions above health and safety, also restricted or stopped cancer investigations and cancer campaigns among vulnerable membership groups, especially those exposed to asbestos (Watterson, 1999b). Some have been captured by the lifestyle approach of charities and the NHS. Others were incorporated by employers. Yet others, including bodies such as the Trade Unions Congress (TUC), appear to have accepted the "Doubt is their Product" arguments of Michaels (2008) but neglected cancer prevention until very recently. They have operated within a tripartite structure with employers and government that has further muted actions on cancer prevention and have yet to mount major campaigns. They run with the lowest cancer estimates in the workplace even though the researchers they cite continue to indicate occupational cancer underreporting (Rushton et al., 2010; TUC, 2012).

Government statistics usually significantly underestimate or downplay occupational cancers; industrial injury schemes rarely recognize them, and insurers under-record them. With the exception of asbestos-related cancers, civil law claims are also few in number. The position in other countries can be radically different: for instance, in Canada, where there has been recognition of various cancers in firefighters and early and rapid recognition of a wider range of cancer sites due to asbestos exposure (O'Neill et al., 2007).

COMPETING INTERESTS AT PLAY? / 147

Government agencies provide data and hence shape policy that does not reflect all the evidence available or delay using data. The HSE has relied too heavily just on data from the European Union and the IARC; although even with the IARC, it may not act. For example, HSE briefings in the latter part of the 2000s on benzene, cadmium, and diesel exhaust were dangerously outdated and substantially underestimated the potential cancer risks (O'Neill et al., 2007: 435). HSE systems to review rapidly and act on new information or revise assessments often left much to be desired (Kellen, Zeegers, Hond, and Buntinx, 2007; Kjaerheim, 1999). Germany and Denmark recognize bladder cancer in the metal industries in which exposures to cadmium and epoxy resins may occur. Denmark officially notes the use of azo dyes by painters; a link not made in the United Kingdom, despite recognizing the hazards to printers. The Sheffield Occupational Health Project (now SOHAS) identified a series of bladder cancer cases in which exposures to cadmium had occurred in situations ranging from smelters to TV repairs and cutlery work (O'Neill et al., 2007).

NIGHT AND SHIFT WORK—A CASE STUDY OF COMPLEX RESPONSES AND SIMPLE NEO-LIBERAL ECONOMICS?

Firefighting

I have been doing nightshift work for over 34 years. I work in a Fire and Rescue control room. I am 55 years of age and I am healthy so far. There are quite a number of other staff who have also worked the same shifts as myself for over 20 years. The shift system we work is the same as the operational firefighters— that is 2 × 10 hour dayshifts and 2 × 14 hour night shifts. Firefighters can retire after 30 years' service or by age 55—whichever comes first. However, control staff work on until age 60 or 65 years. In the future, the operational firefighters will also be expected to work until they are 60 years of age. This translates into working night shifts for 40 years plus, assuming a person begins their career at 18 years old. I believe [this] is unhealthy with unrealistic expectations of a working person. I have been on dayshift since 1 January 2009 and I feel good, I sleep better, eat at regular times, and my stomach isn't complaining anymore. I will have to return to shift work for the rest of my career (five years) and I am concerned because I strongly believe that health is wealth!

As the firefighting box above indicates, the impacts of night shifts are considerable and in current times may mean employees work such shifts for over 40 years.

The case of night and shift work serves to illustrate many of the competing interests and their sometimes complex interactions and subtleties with regard to action to prevent occupational cancers and compensate victims (Watterson, 2009). Around 20% of the workforce in Europe and North America are currently shift workers, with around 3.5 million such workers in the UK, including an estimated

148 / SAFETY OR PROFIT?

400,000 women working nights. In the past, shift work and night work was done primarily by industrial workers in unskilled, semiskilled, and skilled occupations and by a range of emergency workers. The health and safety risks were therefore unevenly distributed, but more white collar workers may now be doing shift and night work, although the directors' car park usually still empties before the start of the night shift, and the tea-time exodus frequently takes with it some and sometimes all of the health and safety advisors and first aiders. UK workers continue to work some of the longest hours in Western Europe (Eurofound, 2011:17). Only workers in Romania and Luxembourg average longer hours, leaving the UK 25th out of 27 European countries in the working hours league table.

Long working hours have been a recognized cause of health problems for centuries. In 1973, shift work was described as "probably the worst of all work patterns" (Kinnersly, 1973: 33), with research showing the adverse effects of "shift lag" and disturbance of the body clock that included fatigue, gastrointestinal effects, and "nervous disorders." Occupational cancer risks were unknown at that time, but standard UK occupational health textbooks in the early 1980s began to link rotating shift work with problems of the endocrine system as well as the circadian rhythms of the body (Harrington and Gill, 1983: 81). This was when alarm bells could have rung about the possible carcinogenicity of nightshift work because it suggests a plausible mechanism existed. Workers who experienced night work could describe the individual effects of such work on themselves and the social disruption caused to them but lacked the sophisticated epidemiological tools to assess specific and longer-term health ill-effects (Williams, 1969: 207, 219, 226–227).

The Danish Government decision to allow compensation for women with occupationally-caused breast cancer discussed later in this chapter, has enormous implications. This is illustrated by the comments of a UK physician in the box below.

Shift work and work at night has now been more firmly linked to a wide range of health problems, including breast cancer, prostate cancer, non-Hodgkin's

---

A Doctor's Note

I'm a consultant anaesthetist who developed breast cancer at the age of 35 after life as a junior doctor doing the old-style 84-hour contracts and 1-in-3 on call rotas with massive sleep deprivation. It's not clear whether female anaesthetists or doctors have a higher incidence than would be expected, but I have four other anaesthetic colleagues who have had pre-menopausal breast cancer without a family history. We have been reflecting on our experiences and risk factors and are very interested that the Danish government has taken this step. It has huge implications for health care workers, and 50% of the current medical graduates are females (Watterson, 2009). Concerns relate not only to those who have had breast cancer prescribed but to a wide range of workers who may be at risk in the future and who do not see any strategies currently adopted that may reduce their risks and ensure effective risk management.

lymphoma, heightened injury risk, heart disease risk factors, and pregnancy problems (Watterson, 2009). Against this background of growing scientific evidence, there are still competing interests denying several ill-health links. Denial of such evidence by employers means that they can continue operating existing shift patterns that benefit production or ensure best use of staff resources rather than prioritizing the protection of the health of their nightshift workers. Hence, employers avoid disruption to existing work patterns and shifts, escape stricter enforcement, and remain free from possible state and legal compensation payments.

The Science

In 2003, in a report on epidemiological studies commissioned by the HSE, the view was that "the possibility that shift work per se increases the risk of breast cancer cannot be dismissed, but on the other hand it remains possible that the apparent associations are due to confounding" (Swerdlow, 2003: 1).

In 2007, the IARC, a World Health Organization agency, announced it would list shift work as a "probable" cause of cancer. The IARC's expert working group comprised 24 experts from 10 countries (Straif, 2007). They drew on two distinct types of information: reports of lab tests that explored mechanisms that might explain how shift work and especially night work might cause cancer; and health surveys or epidemiology papers that were smaller in number but indicated some associations between the cancer and night workers. The weighting given to the mechanisms of carcinogenicity was far higher than that given to epidemiological studies, critical to ensuring a precautionary public health approach. The IARC found shift work involving circadian disruption was probably carcinogenic to humans. Long-term women night workers had higher breast cancer risks than women who did not work nights. The studies involved mainly nurses and flight attendants and were consistent with animal studies showing constant light, dim light at night, or simulated chronic jet lag substantially increased tumor development. Other lab studies showed that reducing melatonin levels at night increases the incidence or growth of tumors. Some theories have also included exposure to electromagnetic fields. Disruption of the circadian system by such means apparently changed sleep-activity patterns, affected melatonin production negatively, and impacted tumor development. Night work proved the most disruptive shiftwork pattern. The excesses of cancers in this group of workers remain. More recent research in Denmark has strengthened the link further, especially for women working in the health sector wherein the most disruptive shifts have been worked (Hansen and Stevens, 2011).

Governments, Their Regulators, and Occupational Disease Compensation Boards

Following the IARC 2002 announcement, by March 2009, some 38 Danish women had been compensated for breast cancer linked to their work wherein the disease existed with no other confounding factors and when 20 or more years had been spent on night shifts (Wise, 2009). They were mostly flight attendants and nurses (CRUK, 2009) but also included some members of the armed forces. The Danish authorities, through their labor inspectorate, then almost immediately focused enforcement activities on hazardous shiftwork schedules, especially the combination of long working hours, night work, and counter-clockwise, as opposed to shift systems that follow the clock, work schedules, measures that will benefit workers facing both safety and health risks from unsatisfactory work patterns.

In the UK, by contrast, when responding to questions on the IARC's breast cancer warning, the HSE said it would take no action until research it had commissioned on breast cancer and night work was complete (Watterson, 2009). This is now expected in 2014. Yet the HSE-backed study does not cover the same ground as the IARC and Danish reviews and begs the question why the HSE could not accept the IARC assessment of carcinogenicity, which was considered authoritative enough to trigger action on both compensation and enforcement by the Danish authorities. The HSE felt able to state in 2009 that "based on the current evidence . . . the principal risk from shift work is fatigue which can contribute to human error, accidents and injuries" (SHP, 2009: 7). A year later, in 2010, an HSE-commissioned study into occupational cancer indicated that nearly 2,000 women a year could contract breast cancer occupationally due to shift work (Rushton et al., 2010: 56). The response from the HSE to this research remains minimal. The competing interest of their own commissioned research with governmental deregulatory policies and plans to get sick people back to work when work does not exist appears to have anaesthetized them. Senior HSE inspectors have indicated that they recognize occupational disease is a UK priority, but they have to react all the time to "safety enquiries," and no one raises occupational cancer as an issue. This presents the HSE at the top as a passive body unable to act on its own in generating action on serious workplace health threats. Other field inspectors have stated that they lack the knowledge and expertise to deal with occupational diseases beyond simple dermatitis cases and so do not pursue matters (A. Watterson, personal communications, 2009–2010).

Consecutive UK governments have opposed full implementation of the Working Time Directive and restricting hours. The HSE proved almost dormant on the issue. Since 2001/2002, the HSE has issued 38 improvement notices relating to the Working Time Regulations, although since 2003, no year has seen more than two notices. The two HSE prosecutions since 2001 were not on working-hours-related health issues, but followed workplace fatalities. In the period from 2004/2005 to 2007/2008, the HSE managed just 16 prosecutions in the UK on any offenses related to work-related ill-health. The HSE now visits the average workplace less than once every 14 years, and the night shift is almost entirely beyond official scrutiny too. The HSE's website makes plain that, unlike the health service or other criminal law enforcers like the police, normal service is not available outside of office hours: "In our experience, most incidents may be satisfactorily investigated during normal working hours," (HSE, 2009).

Risk assessment and related risk management should operate under EU laws on risk assessment and the EU Working Time Directive despite its many derogations in the UK. These laws, limits, and health assessments (if a night worker) are enforced by the HSE, local authority environmental health departments, the Civil Aviation Authority (CAA), the Vehicle and Operator Services Agency (VOSA), and the Office of Rail Regulation (ORR). The working time regulations state that an "adult worker must have the opportunity for a free health assessment before being assigned night work with regular re-assessments later" (HSE, 2013). But how "opportunity" is interpreted and who carries out the assessments and reassessments is not specified.

Few workers across the UK have access to effective occupational health services. In 2006, the Department of Work and Pensions reported that "only 15% of firms provided even basic occupational health support, with only 3% providing comprehensive support" (Pickvance and O'Neill, 2007). There has been no recent dramatic change in the position in a country that still does not ratify the ILO Convention on Occupational Health Services. With such low occupational health coverage and such widespread use of shift work and night work, there have to be serious questions about the effectiveness of any regular health checks for night workers and shift workers. It also raises the further question of how well employers are capable of conducting effective and continuing health surveillance of vulnerable shift workers and night workers.

The HSE has published "good practice" pointers on shift working, and its general online shiftwork guidance noted several "key risk factors in shift schedule design" that should be considered when assessing and managing the risks of shift work. These included workload, work activity, shift timing and duration, direction of rotation, and the number and length of breaks during and between shifts (HSE, 2006). They also recognized that other features of the workplace environment such as the physical environment, management issues, and employee welfare could contribute to shiftwork risks but did not mention cancer risks. However, the HSE did not appear to enforce these issues because the few enforcement notices that have been issued were "principally in relation to failure to provide free medical assessments or in relation to record keeping" (Hazards, 2009).

The Trade Unions and Employees

The TUC biennial surveys of safety reps regularly flagged long hours as a top health and safety concern. In 2008, the Australian Workers Union called for a review of working hours after the IARC announcement. The Australian New South Wales Nurses Association launched a claim in the Industrial Relations Commission in May 2009 linking shift work with higher rates of breast cancer, heart disease, miscarriage, clinical depression, and divorce. But many of the unions in the UK have not campaigned either loudly or at all for government to adopt the Danish compensation and enforcement positions on night work linked to cancer.

The TUC works in a tripartite setting but has publicized Danish action and raised the issue of breast cancer linked to nightshift work with the HSE. Individual trade unions with members who work shifts have different responses, but few have mounted campaigns and calls for policy changes on the subject. The British Medical

152 / SAFETY OR PROFIT?

Association has no formal OHS advisor and no position on nightwork health risks. Unite is active on the issue and has members who are air crew or health and emergency workers, for example. Remarkably, the Royal College of Nurses appears to have no position either on the matter nor does the Prison Officers Association. The Fire Brigades Union is active on the wider issue of occupational cancer but not specifically on the links between cancer and night work.

Individual National Union of Journalists' members have shown a greater interest in the topic triggered by their own and their colleagues' working hours, which have often included many years of shift and night work. Individual workers in the UK have also flagged their concerns about night work as a carcinogen. Such well-informed employees were sometimes members of those trade unions that also operated as professional bodies and yet lacked full-time health and safety advisors.

The Employers

Employers' groups, like trade unions, have many and varied responses to occupational cancer prevention. Larger employers may have occupational health staff; smaller ones will usually not. Larger employers may have greater flexibility in dealing with health and safety and have choices about which shift patterns to select and what facilities to offer for shift and night workers. Smaller employers may not. However, the lack of occupational health services across all UK employment sectors indicates not only changes in economic and employment structures but also a lack of commitment to occupational health. The CBI has, for example, no explicit reference either to occupational cancers or the health effects of night work on its main web pages but does flag costs to the economy of workplace absence (CBI, 2012). In contrast, the UK Engineering Employers Federation (EEF), in 2009, did urge employers not to be complacent about occupational cancers and urged some firms to do more to protect employees (Baker, 2009). The EEF also provided general information on their web pages about shift working and health and safety. The Federation of Small Businesses (FSB) noted that very few small and medium enterprises had occupational health services, and there is no specific information available to their members through their web site on the hazards of shift and night work (FSB, 2012). However, several small business spokespersons quickly and loudly dismissed the work of the IARC and the actions of Danish agencies in 2009 for compensating night workers for occupationally caused breast cancer. No scientific evidence was provided to refute the IARC work (Watterson, 2009).

Employers have resisted many provisions of the Working Time Directive, and it is clear that many fail to offer either health assessments or proper occupational health services to night and shift workers. Many, but not all, employer bodies also support deregulatory action on UK health and safety laws and contested the findings of the IARC on night work and cancer links drawing not on science but politics to support their position. The HSE's inertia in enforcing the laws relating to risk assessments and health checks is by default a position favorable to employers and governments who espouse a deregulatory ideology. It enables a "don't look, don't find, no need to act" philosophy, which is increasingly dominant. This presents the HSE at the top as a passive body unable to act on its own in generating action on serious workplace health threats.

A range of industry responses emerged through research on the risks of nightshift work. These included denial of the validity of positive studies, emphasis on the validity of negative studies, support for the HSE in delaying any action, support for downplaying of the need for HSE inspections and enforcement linked to burdens on industry and deregulatory arguments, and finally support for a UK opt-out on the Working Time Directive. Some law firms also tried to obtain business with suggestions that women night workers would be fired because of potential risks, and hence lawyers would be needed in tribunal cases.

Civil Society Organizations

Some NGOs—such as Hazards, linked to both the TUC and activist groups, and the Alliance for Cancer Prevention, which brought together environmentalists, worker groups and trade union officials—repeatedly highlighted the cancer risks of night work. They also called for effective enforcement of existing laws and compensation along the lines of the Danish action. This was geared to addressing some of the threats indicated by the Danish reviews and the IARC data (Watterson, 2009). They advocated the precautionary principle in viewing new data and a public health response informed by a balance of probabilities approach, which failed safety and not danger. They emphasized complex interactions and cumulative effects on health along with calls for compensation of women night workers along the lines of Danish Government actions and the IARC listing of shift work as a 2a carcinogen.

NGOs that came from the charity sector and stressed medicalized cancer treatment rather than a cancer prevention strategy usually adopted a different position. They dismissed the links between breast cancer and nights. They focused primarily on investigating cancer mechanisms and research on the treatment of the diseases. Key "prevention" work has historically come in two forms. First, there is the misnamed "secondary prevention," which relates to early diagnosis and treatment, not prevention. More and earlier screening for breast cancer is now being mooted by some as a solution to the hazards of shift work for women. Second, such charities adopt a strong, often almost exclusive, media emphasis on lifestyle factors and voluntary risks involved in cancer causation—diet, alcohol, exposure to sun—rather than the involuntary risks revolving around, for example, life circumstances or environmental and occupational exposures to carcinogens. Organizations such as Breakthrough Breast Cancer have been publicly quoted on the Danish night worker and breast cancer link as stating that "Whatever their working hours, all women can help reduce their risk of developing the disease by maintaining a healthy weight, taking regular exercise, and limiting the amount of alcohol they drink" (SHP, 2009: 7). Yet research indicates that nightshift work is itself physiologically a cause of weight gain as well as a factor in exposure to junk foods through lack of access to fresh food in canteens that were closed at night (Watterson, 2009).

154 / SAFETY OR PROFIT?

Night Work and Cancer: An Overview

Competing interests exist with regard to calls to prevent breast cancer in women night workers and to compensate those who are affected. In Denmark, the arguments of the IARC and the compensation boards have prevailed, supported by actions from their enforcement agencies and further validated by scientific studies. In the UK, the neo-liberal agenda has prevailed, and the HSE was captured by that agenda. Hence, the HSE ignored its own scientists' reports. Bolstered by the cancer charity lobby and effectively unchallenged by many of the trade unions, it was able to deny international agency evidence and has continued to be inactive on breast cancer and night work. The strength of the neo-liberal agenda easily disarmed media interest in the subject. The case of the relation between night work and cancer is not unique and in many respects reflects the wider responses of the UK government and industry to calls for cancer prevention. Chart 7.3 illustrates the key general approaches used. Column 1 shows UK government action, Column 2 shows drivers for media interest and coverage, and Column 3 shows Danish state action.

CONCLUSIONS

The UK currently lacks any effective strategy on occupational cancer prevention and has done so for over 30 years. This is despite the existence of substantial scientific research on the subject. Examples from France, the United States, Canada, and Denmark have shown how such a strategy could be developed and then how it might benefit employers, employees, and government alike. These benefits for industry and employers have been demonstrated by bodies such as the Toxics Use Reduction Institute (TURI) in the United States. TURI showed over 20 years that removing and/or reducing carcinogen use had economic benefits for employers in reduced expenditure on materials and storage costs, worker health and safety improvements, and reductions in environmental pollution, while boosting production and retaining advantage in technological innovations (Eliason, 2008). Governments would benefit from such cancer prevention strategies through reduced healthcare costs and potentially better economic growth as fewer employees fall ill. The reasons for inaction lie in industry hostility toward proposals that challenged employers' rights to manage and the raft of related neoliberal policies. Researchers are rarely funded to explore prevention options and pursue the research agenda of government and industry. Trade unions fitfully (and sometimes not at all) pursue a limited occupational cancer prevention agenda, although their members often show greater interest in the subject. Cancer charities usually neglect cancer prevention, often gain funds and support from industries that produce carcinogens, and focus almost exclusively on individual lifestyles rather than life circumstances, which include work and wider environmental factors.

Removing carcinogens and reducing exposures to carcinogens has been more widely mooted in recent years by governments and international agencies than at any

COMPETING INTERESTS AT PLAY? / 155

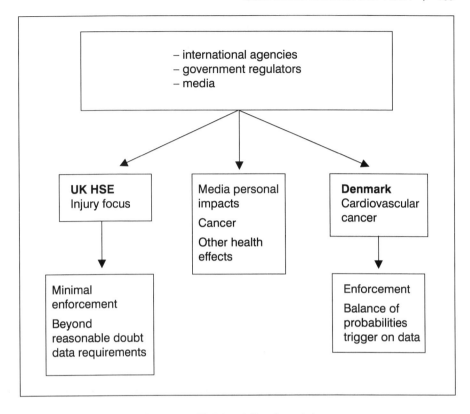

Chart 7.3 Weighted filtration of data.

time since the early 1980s (Christiani, 2011; Hazards, 2009; Watterson, 2012). U.S. states have successful toxics-use reduction programs that cut carcinogen usage (TURI, 2012), and the U.S. president's recent commission on cancer prioritized cancer prevention. France has an active national occupational cancer prevention strategy, as does Canada. The WHO-sponsored conference in Asturia, Spain, on the subject identified the key practical steps needed (Landrigan et al., 2011) which were that

- The WHO should develop a global framework for control of environmental and occupational carcinogens, which concentrates on the exposures identified by the IARC as proven or probable causes of human cancer (IARC, 2011).
- The WHO should develop measurable indicators of carcinogen exposure and cancer burden to guide cancer surveillance worldwide.
- All countries need to adopt and enforce legislation and regulations to protect their populations, especially the most vulnerable (pregnant women, fetuses, infants, children, and workers) against environmental and occupational cancers.

- All countries need to develop communication campaigns tailored to local needs in order to educate their populations about environmental causes of cancer and prevention strategies.
- Corporations should comply with all rules and regulations for prevention of environmental and occupational cancers and adhere to the same standards in all countries—developed and developing—in which they and their subsidiaries operate.
- The UK should be able to use such a declaration to overcome a long history of unequal competing interests and move forward quickly in developing an effective cancer prevention strategy.

Time will tell.

http://dx.doi.org/10.2190/SOPC8

CHAPTER 8

The Limits and Possibilities of the Structures and Procedures for Health and Safety Regulation in Ontario, Canada[1]

Wayne Lewchuk

INTRODUCTION

In 1972, the Robens Report, *Safety and Health at Work*, set the tone for health and safety debates for several decades in the UK and in a number of English-speaking countries (Walters, 2003). High rates of injury at work were explained by the prevalence of employer and worker "apathy" toward health and safety questions. One year later, Theo Nichols and Peter Armstrong wrote a penetrating critique of the Robens Report (Nichols and Armstrong, 1973), warning that it was based on flawed assumptions and that regulations based on these assumptions would leave "millions of men and women" exposed to unacceptable levels of workplace risks. With clarity of analysis that became the hallmark of his writing, Nichols was critical of the superficial analysis upon which the apathy argument was based. He argued that it was relatively easy to attribute accidents to factors such as "lack of attention," "inadequate supervision," or "worker's fault," but that behind most

---

[1] I would like to thank Scott Thorn, who helped assemble many of the briefs that this chapter is based on. I would also like to thank the members of the Labour OHCOW Academic Research Collaboration for sharing their extensive knowledge into the inner workings of health and safety regulation in Ontario and to Joan Eakin for comments on an earlier version. LOARC made its own presentation to the panel titled, Internal Responsibility: The Challenge and the Crisis, August 25, 2010.

accidents was a more fundamental explanation. He stressed that Robens was explaining accident rates without seeing them in the context of the "social relations of production." According to Nichols and Armstrong, most accidents occurred due to the pressure workers were under to keep production going. Accidents had more to do with the drive for profits than workplace apathy.

In the years immediately following the Robens Report (1972), both Britain and Canada would adopt new regulatory "structures and procedures" that emphasized self-regulation by workplace parties at the expense of external enforcement of health and safety standards by inspectors. Such structures and procedures were intended to increase employer and worker sensitivity to safety issues in the workplace and to overcome the prevalence of workplace apathy. Research over the last 20 years has explored the efficacy of these changes (Hall et al., 2006; Levesque, 1995; Lewchuk et al., 1996; O'Grady, 2000; Tuohy and Simard, 1993; Walters, 2003). This chapter has a more modest goal: to explore labor, management, and government concerns regarding the health and safety regulations and how these need to change in light of the spread of precarious employment.

In January 2010, the Province of Ontario announced the formation of a panel to review the effectiveness of health and safety regulation in the province. The numerous submissions by employer groups, unions, and other interested organizations provide a rare window into different views on how well the system is functioning and the changes that are needed to adapt to the spread of precarious employment. The Review Panel's final report, released in December of 2010, provides insight into the future direction of health and safety regulation in Ontario and how one province proposes to deal with the concerns raised by the spread of precarious employment. It also speaks to the larger issue of how labor and employers try to shape government decisions and how the government, in defining the scope of a review and the resources available to it, can significantly influence its recommendations.

THE ROBENS REPORT AND THE MOVE
TO SELF-REGULATION

The Robens Report (1972) took the view that "prescriptive" regulatory systems that relied on standards codified in law, enforced by an independent inspectorate, and backed by the threat of fines and/or criminal prosecutions for violations were not making workplaces safe. The problem, according to Robens, was general "apathy" by employers and workers toward health and safety issues at the workplace, an apathy that even a well-resourced inspectorate could not surmount. At the time of the Robens Report, HM Chief Inspector of Factories, Mr. Brian Harvey, argued, "However many inspectors one has, one cannot take the responsibility away from industry of putting its own house in order" (Nichols and Armstrong, 1973: 37). Robens advocated a shift to a radically different regulatory system that emphasized workplace-based self-regulation of occupational health and safety by employers and workers. The goal was to develop "structures and procedures" at work to get employers to do the "right" thing with only a limited threat of external enforcement and even then, only in the most egregious cases.

The 1972 Robens Report led to the UK Health and Safety at Work Act (1974). As proposed by Robens, there was a shift away from a prescriptive regulatory system and a move to structures and procedures that encouraged employers to self-manage workplace health and safety risks. As pointed out by others, "Such provisions have a constitutive and structuring function for employers and they require them to focus on the organizational means with which they are equipped to assess and manage risks" (Walters, 2003: 3). Increasing employer and worker responsibility for managing their own safety at work was viewed as a remedy for the apathy that Robens argued was the cause of many accidents.

ONTARIO AND THE
INTERNAL RESPONSIBILITY SYSTEM

In 1974, the province of Ontario[2] appointed the Ham Commission to investigate the causes of workplace injury in the province. The Ham Report was issued in 1976 and, like Robens, Ham raised doubts about the efficacy of a regulatory system that relied heavily on standards codified in laws and enforcement by an external inspectorate. For Ham, the key weakness of the existing regulatory framework was the ineffectiveness of an underdeveloped "internal responsibility system" resulting from the inability of labor to voice its concerns and to contribute to making workplaces safer. Ham argued, "The worker as an individual and workers collectively in labour unions or otherwise have been denied effective participation in tackling these problems; thus the principles of openness and natural justice have not received adequate expression" (1976: 6).

Ham proposed shifting more responsibility for workplace safety to the internal responsibility system and making this system more effective by giving workers more voice in health and safety matters. Employers would be encouraged to self-regulate, and workers would be empowered to ensure this happened. Worker-auditors, modeled on the system in place in Sweden, were to be appointed, and joint health and safety committees were to become mandatory in all mines and factories. Ham's proposal could be characterized as a move to a collective laissez-faire regime as discussed by Tucker in this volume.

Ham's recommendations became the basis for the 1978 Ontario Occupational Health and Safety Act.[3] As was the case in the UK, a new regulatory regime was adopted that placed greater emphasis on self-regulation by employers and workers, and a reduced emphasis on external regulation through inspectors and legal restraints.

[2] Occupational health and safety in Canada is predominantly a provincial jurisdiction.

[3] See Tucker, this volume, for details.

160 / SAFETY OR PROFIT?

ONTARIO REVIEWS OCCUPATIONAL HEALTH
AND SAFETY REGULATIONS:
THE DEAN REVIEW PANEL

More than 3 decades later, the province of Ontario appointed a review panel to assess the effectiveness of its occupational health and safety regulatory framework. Much has changed in the Canadian economy in the ensuing decades. Most Canadian companies face more intense competition because of the free-trade agreements that are opening Canadian markets to Mexican and American companies. With trade liberalization, Canada's economy has become more export oriented. At the same time, union density has fallen, and changes in legislation and workplace organization make it more difficult to organize and represent workers. Labor markets have also changed. There has been a growth in nonstandard employment and in particular, self-employment, temporary agency employment, and other forms of precarious employment. The workforce has become more diverse, and a growing segment of the labor market is made up of migrant workers. Many firms have reduced direct employment and rely more heavily on extended supply chains. One result of this has been a proliferation of small employers.

There are indications that provincial officials were already aware that self-regulation was leaving many Ontario workers exposed to unacceptable levels of risk. In 1998, they announced a Prevention Strategy "to develop greater communication and collaboration" among the agencies involved in occupational health and safety. Perhaps more telling was a limited shift back to external enforcement of health and safety standards. The Ministry of Labour implemented a "safety blitz" inspection strategy, "designed to raise awareness and increase compliance with health and safety legislation." This was accompanied by a growing focus on poor health and safety performers, which included more frequent inspections of the 10% of firms with the worst safety records and more direct assistance from the relevant health and safety association charged with educating employers and workers in these sectors. Between 1970 and 2000, the number of workplace inspections had declined over 50%, despite the growth in the economy. This trend was reversed starting in 2002. By 2007, the number of workplace inspections had doubled (Ontario Ministry of Labour, 2011; Tucker, 2003).

It is instructive to review the immediate events that led to both the 1974 Ham Commission and the 2010 Dean Review. They are symbolic of the larger changes that have taken place in the Ontario economy over the last 3 decades. In the early 1970s, the province's decision to enact new health and safety legislation was strongly influenced by the demands of unionized workers in the mining and industrial sector. A 1974 wildcat strike by miners in northern Ontario, who were represented by the United Steel Workers, was the immediate event that precipitated the Ham Commission. The workers calling for reform in the early 1970s were predominantly high seniority, white male workers, represented by unions at the peak of their postwar power. These workers enjoyed above average pay and benefits and were demanding improved health and safety conditions at work (Storey, 2005). The influx of immigrant labor into the labor market was only beginning, and only a small number were members of the unions lobbying the government for new regulations.

The Dean Review was announced under very different circumstances. While the government had been moving toward launching a review for some time, the announcement of the review came shortly after a tragic Christmas Eve 2009 accident, which saw four workers plunge 13 floors to their deaths when the scaffolding they were using at a building restoration site failed. The conditions that led to this tragedy are under review at the time of writing, but the broad facts of the event seem to be generally agreed upon. The workers involved were unorganized immigrant workers, employed as subcontractors. Those on the failed scaffolding had been on the job only a few months. It is unclear what, if any, training they'd received. The one experienced worker on the job, who got off the scaffolding moments before its collapse, suggested a better understanding of the job might have saved lives. It is claimed there were only two safety harnesses between the five to six workers on the job, and those wearing them were not properly attached to a lifeline. According to the wives of two of the victims, "Their husbands were pushed to get as much work done as possible on the high-rise before breaking for the holidays."[4]

While the Christmas Eve event may have played a small role in the decision to set up a review panel, it played a major role in focusing attention on the issue of changing labor-market conditions and the ability of the existing regulatory system to protect the health of workers in precarious employment relationships. As an article in the local newspaper asked, "Are rogue companies using shady methods to attract illegal workers who won't, or can't, assert their rights? Are they improperly training workers or skirting safety laws designed to protect them?"[5] In essence, is profit taking priority over safety?

In January 2010, the Review Panel was charged with examining the continuum of safety practices in a workplace, including entry-level safety training, the impact of the underground economy on health and safety practices, and how existing legislation serves worker safety.[6] The underground economy refers to a class of firm that operates outside of the standard regulatory regime for construction firms. Most are small scale firms doing subcontracting work with no or limited written contracts). Four months later, following the fallout from the events on Christmas Eve, its mandate was defined to include

- the roles and responsibilities of the system partners,
- the impact of the underground economy on workplace health and safety,
- the protection of vulnerable workers,

[4] Toronto Sun, December 10, 2010. http://www.torontosun.com/news/torontoandgta/2009/12/27/12276031-sun.html. See also Building and Concrete Restoration Association of Ontario (2010) See footnote 9 for the reference to the Building and Concrete..

[5] Toronto Star, January 10, 2010. http://www.thestar.com/opinion/editorials/article/748382—probe-scaffold-accident

[6] Ministry of Labour, January 27, 2010.

162 / SAFETY OR PROFIT?

- the use of incentives to motivate superior health and safety performance,
- linking procurement of goods and services to health and safety performance,
- the role of joint health and safety committees,
- the impact of advancements in technology/innovation on health and safety, and
- mandatory entry-level health and safety training.[7]

While we cannot know for sure what was in the mind of Ham in the mid-1970s when studying occupational health and safety regulation, it is unlikely that issues of vulnerable workers, precarious employment, the underground economy, or supply chains were foremost. Making these key issues in such a high-level review marks an important phase in the evolution of workplace health and safety regulation. At the same time, the ability of the Panel to seriously deal with these issues was limited. The Panel's mandate was to make the existing system more efficient rather than propose radical reforms. Hence, if the issues raised by precarious employment were to be dealt with, they would have to be dealt with by fitting a round peg into a square hole. The Panel had neither the time nor the resources to do more.

PRESENTATIONS TO THE REVIEW PANEL

To deepen our understanding of the concerns regarding the existing regulatory framework and to assess views on how these regulations need to adapt in the face of the spread of precarious employment, it is useful to look at the presentations to the Panel by employers and organizations representing workers. I have relied on the presentations of major organizations that made their reports available on the Internet during the review process. I focus on the presentation by three employer groups and three organizations representing labor. The Panel received over 100 formal presentations. The six that I focus on represent major players in the debate over health and safety and can reasonably be expected to reflect the overall tone of what the Panel heard.[8] The three employer organizations are the Building and Concrete Restoration Association of Ontario, whose members are involved in the

---

[7] Ministry of Labour, Consultation Paper, April 2010.

[8] Other significant commentaries on the Dean Review (2010).

1. Canadian Union of Public Employees, Submission to the Expert Advisory Palen on Ontario's Occupational Health and Safety Public Consultation, June 28, 2010.

2. Marentette, Rolly, Chair, Windsor and District Labout Council Health and Safety Committee, Submission to Health and Safety Act Review Panel, June 4, 2010.

3. National Union of Public and General Employees, No Ontario plans for health and safety privatization. http://www.nupge.ca/content/3283/no-ontario-plans-health-and-safety-privatization

4. Ontario Association of Non-Profit Homes and Services for Seniors, June 30, 2010. http://www.oanhss.org/AM/AMTemplate.cfm?Section=HomeandTEMPLATE=/CM/ContentDisplay.cfmandCONTENTID=6782

5. Ontario Chamber of Commerce, OCC Working to Create an Improved Health and Safety System for Ontario, December 7, 2010. http://occ.on.ca/2010/12/occ-working-to-create-an-improved-health-and-safety-system-for-business/

type of work associated with the Christmas Eve accident; the Business Council on Occupational Health and Safety (BCOHS), an umbrella organization that counts many of Canada's largest employers as members; and the Ontario Chamber of Commerce. On the labor side, I included the Toronto Workers' Health and Safety Legal Clinic, who provide legal advice and representation to nonunionized low-wage workers who face health and safety problems at work; the Ontario Public Service Employees Union (OPSEU), the major union representing public-sector workers; and the Ontario Nurses' Association (ONA), who provided a comprehensive brief to the panel.

EMPLOYER BRIEFS TO THE REVIEW PANEL

Building and Concrete Restoration Association of Ontario (B&CRAO)[9]

The brief from the Building and Concrete Restoration Association of Ontario (B&CRAO), whose members use suspended access equipment (SAE), the type of equipment that failed in the Christmas Eve accident, focused on how to avoid such accidents in the future. A significant theme in their brief was the need for more external monitoring of hazardous work practices such as SAE work, greater regulation of the underground economy, and better training for workers and supervisors. In the introduction to their brief, they refer specifically to the Christmas Eve accident, arguing failure of such equipment should never result in deaths. They invoke Robens' apathy argument, placing blame on the failure of workers and employers to follow accepted safety practices, including the proper securing of fall-protection safety devices to a secure lifeline. They focused on the need for better training, supervision, legislation, and enforcement to avoid such accidents in the future. The social relations of production that led to failure in all of these areas are not explored in any detail nor is there any reference to the claims of the victims' families that the workers were being pressured to complete tasks before the Christmas break.

Many of their recommendations focused on failures in mandated regulations and confusion caused by the language used in legislation and regulations. They proposed seven regulatory changes, including a proposal to mandate notification of SAE projects to the Ministry of Labour; mandatory training of SAE workers, riggers, and supervisors; mandatory posting of work plans; and mandatory posting of SAE system capacity. They also note the lack of regulations requiring inspection of all SAE setups and propose a new regulation requiring an inspection by a "competent worker."

---

[9] Building and Concrete Restoration Association of Ontario (B&CRAO), Recommend Changes to the Occupational Health and Safety System: A Response to the Expert Advisory Panel on Occupational Health and Safety, June 28, 2010. http://www.bcrao.com/pdf/2010/B&CRAO%20-%20Brief%20June%2028%202010%20-%20Final.pdf

164 / SAFETY OR PROFIT?

An important focus of their report was the distinction between legitimate employers operating in full compliance of existing regulations and employers in the underground economy who cut health and safety corners. They suggest,

> Legitimate employers have a stake in compliance with the law. They should be encouraged to show that being safe and legitimate is a profitable way to do business through mentoring programs. Educational programs and safety seminars organized by reputable voluntary associations could contribute to helping motivate employers to comply with the law (B&CRAO, 2010: 11).

To address the spread of underground employers and vulnerable workers, they proposed setting up a hotline number to allow workers to report underground employers. They noted this was not likely to be very effective for vulnerable workers due to the insecurity of their employment. They suggested the Ministry provide more health and safety training at all levels and provide this material in multiple languages. One proposal would require the Workplace Safety and Insurance Board (WSIB) to eliminate the "independent" contractor category. This would result in closer supervision of all jobs and better training of workers. Independent contractors are generally not governed by a collective agreement, own their own equipment, and provide services to more than one company. Implicit in the Association's brief is a concern that some companies are putting profits before safety by using the independent contractor option.

A key proposal was to extend responsibility for monitoring health and safety to building owners who currently have no role in this process and, it is argued, may be putting contract cost before safety. This can lead to building owners employing unqualified contractors and ignoring safety issues. Their proposals included requiring owners to stress safety as one component of new contracts and to provide contractors with specific safety guidelines. They suggested requiring building owners to conduct safety audits and employ third-party safety inspectors as well as submitting documents that verify the contractor has supplied and submitted safety documents to both the Ministry of Labour and the WSIB. This can be viewed as an extension of the self-regulation approach to health and safety. Somewhat surprising were recommendations directly related to gaps in the self-regulation approach to health and safety. Three decades after the shift to self-regulation, there was still the need, in the words of the Association, for

> a simple flow chart outlining the responsibilities . . . from the perspective of an employee and an employer and how they are inter-connected. We urge the use of icons and language that is easy to understand to help overcome the language barriers that some workers face (B&CRAO, 2010: 4).

In their view, self-regulation was not enough. They called on the Ministry of Labour to employ more inspectors and to provide better training of inspectors. The B&CRAO brief explicitly recognizes the profit versus safety trade-off, but the focus is on rogue companies operating in the underground economy and not the reputable

STRUCTURES FOR REGULATIONS IN ONTARIO, CANADA / 165

firms represented by B&CRAO. Their proposals can be viewed as an effort to remove the cost advantage of operating in the underground economy.

Business Council on Occupational Health and Safety (BCOHS)[10]

The overall tone of the BCOHS brief was a call for more cost-effective delivery of health and safety services. They were concerned with overlap by the different agencies and were opposed to measures that might add new costs to the system, including a new prevention agency, new administrative financial penalties, or new regulations to deal with the underground economy. They went as far as to question the actual contribution of the province's health and safety system and called on the government to better demonstrate the system's contribution to improved health and safety. In their view, "Employers are the ones who invest in health and safety, build it into the organization's culture, and put policies, procedures and programs in place" (BCOHS, 2010: 2). Employers should receive more recognition of their efforts. The BCOHS stressed the need for "evidence based analysis" and requested that only solutions that the government can demonstrate are effective be implemented.

The BCOHS's brief focused on the duplication of services and the overlap between the enforcement functions of the Ministry of Labour and the WSIB and the prevention roles of the health and safety associations and the WSIB. They expressed concern regarding duplication as a result of having multiple organizations involved in training. As examples of this duplication, they made a special reference to the existence of a separate worker's health and safety center and a worker-administered training system. Their obvious preference was for a standardized training system to which worker-based training would have to conform. The BCOHS also questioned the value of the Occupational Health and Safety Council of Ontario.

Similar to the B&CRAO, the BCOHS points a finger at the failure of health and safety initiatives to reach "poor" performers and "those operating outside of the law." In their view, "Too much of the current System's focus is on those who are already investing in health and safety" (BCOHS, 2010: 2). The underground economy was seen as a problem because it imposed costs on "legitimate" employers. There was less emphasis on the costs it imposed on workers in precarious employment. In the BCOHS opinion, there was no need for new regulatory policies to deal with the underground economy. They wrote,

> Government has the enforcement tools it needs to address the underground economy. Employers who operate legitimately do not want to incur additional costs for enforcement or be penalized in any other way for new efforts by government to address the underground economy (BCOHS, 2010: 7).

---

[10] Business Council on Occupational Health and Safety (BCOHS), Letter to Tony Dean, June 28, 2010: http://www.pcac.ca/LinkClick.aspx?fileticket=-AB0bYwB99Y%3D&tabid=552&mid=1105

They called for more study of why employers chose to operate in this manner before implementing any changes.

While voicing their support for both Joint Health and Safety Committees and the Internal Responsibility System (IRS), they expressed concern about the effectiveness of current certification training that, in their view, was too generic. The IRS was viewed as a useful construct, but one that remains underdeveloped, without clear lines of responsibility or the tools needed to be truly effective.

The BCOHS was opposed to penalties for poor performance, but in favor of rewards for good performance. They called on the government to lead by example and require specific health and safety standards in their dealings with suppliers. Training was also viewed as making an important contribution, but training needs to be more sector specific and delivered at a lower cost than currently provided by the Health and Safety Associations. Training needs to be done in multiple languages, and the efforts of small and medium-sized firms to deliver training through informal on-the-job support needs to be better recognized.

Ontario Chamber of Commerce[11]

The Ontario Chamber of Commerce (OCC) worked closely with other employers in drafting their brief, including voicing support for the views put forward by the BCOHS. They wrote that their "members are fully supportive of any changes that will promote a realistically safer working environment at a lower cost to employers." The Chamber viewed education as the single most important factor in improving workplace health and safety. As was the case with the BCOHS, the preference was for a smaller health and safety regulatory organization at reduced cost to employers. They were supportive of streamlining the system's prevention functions and supportive of the proposal to establish a new Health and Safety Prevention Authority that would bring all prevention activities under one roof. They did express concern that training might become more generic and less sector specific under the new Authority. They generally rejected the use of administrative monetary penalties for violations of safety regulations, but did accept that they might be useful in cases of elevated risk and when penalties could be shown to have a demonstrable impact on compliance. They expressed concern with delays in receiving information that would allow OCC members to improve safety at work. They complained that employers are issued notices by inspectors despite attempts to get the information they need to bring their systems up to the required standards.

There are several common themes running through the employer briefs. The first was their brevity: most were only a few pages long. Second, was the paucity of evidence. Despite their demand that the panel act based on evidence, the employers did not feel the same need in making their own recommendations. The employer briefs focused on reducing inefficiency and overlap in prevention and enforcement

---

[11] Ontario Chamber of Commerce, New Health and Safety Prevention Authority, December 2, 2010.

activities. The overall impression was that employers felt the system had become too expensive and could deliver the same services at less cost. There was a strong preference for evidence-based policies and for rewards for good performance and less enthusiasm for penalties for poor performers. Giving employers more credit for their contributions was stressed and the need for employers to have a larger voice in joint efforts to reform health and safety regulations. Neither the BCOHS nor the OCC put much emphasis on changing labor markets or the problems facing precarious workers beyond the need to better control the underground economy. These proposals appear to have more to do with saving "legitimate" employers costs created by employers operating outside of the mainstream. The BCOHS was of the view that the government already had the tools to deal with the underground economy and vulnerable workers. Only the B&CRAO called for major new initiatives in the face of the growth in the underground economy and vulnerable workers. They accepted that it was unlikely that any changes to the internal responsibility system would improve conditions and hence called for mandatory regulations and more inspectors and closer supervision of employers employing vulnerable workers.

LABOR SUBMISSIONS

Toronto Workers' Health and Safety Legal Clinic (TWHSLC)[12]

The Toronto Workers' Health and Safety Legal Clinic is funded by Legal Aid Ontario and provides legal advice and representation to nonunionized low-wage workers facing health and safety problems at work. Their submission focused on the "vulnerable worker," a worker they define as generally uninformed about their rights, who faces a level of insecurity and desperation in their employment, and who is reluctant to act on job dangers if this might endanger future employment opportunities. Vulnerable workers were described as workers who "enter the workplace as the prey of any employer that seeks to maximise returns through pressuring employees" (TWHSLC, 2010: 3). Unlike the employer briefs, the Clinic's brief did not view vulnerable workers as mainly a problem in the underground economy. The concern was more widespread. They called for major reforms to deal with this issue. The Clinic argued,

> The system that we ask you to abandon is the one where Ministry of Labour leaves health and safety issues in non-union workplaces to the parties to sort out. This antiquated notion expects the most vulnerable to rely on strength that they do not have. We advocate for increased involvement by the Ministry of Labour by being proactive (TWHSLC, 2010: 8).

---

[12] Toronto Workers' Health and Safety Legal Clinic, Submissions to the Expert Advisory Panel on Occupational Health and Safety: Proposed Changes to the Prevention and Enforcement System, June 30, 2010: http://www.workers-safety.ca/remository?do=view&file=publications%3A+newsletter%2C+Workers+Guide%2C+FACT+SHEETS%2C+reports%2C+etc|2010+06|2010+06+30.+Submissions+to+MoL+Panel.doc

168 / SAFETY OR PROFIT?

In their view, the existing system relies too heavily on the goodwill of the employer to voluntarily select injury prevention over profitability. They see no way to make this system work for vulnerable workers. They called for a new right: the "Right to a Government Inspected Workplace." This would include increased monitoring by Ministry of Labour inspectors and greater deterrence through increased fines for noncompliance. While not ignoring the role of better training, they pointed out that health and safety in Ontario is not a "battle against ignorance" but rather the failure of the complaints-driven process in place and the need for a proactive enforcement model.

In addition, they envisioned a workplace safety posting system, including posting the date of the last inspection; whether the workplace received a pass, fail, or conditional pass rating; and 24-hour access to a health and safety desk to direct inquiries. They noted that a posting system is already in place for restaurants in many jurisdictions across Ontario. External enforcement was seen as the key to improved health and safety for workers in precarious employment. They called for an increased presence and increased authority for inspectors and a willingness to prosecute employers charged with reprisals against workers who raise health and safety issues. Inspectors should be mandated to force employers to reinstate workers who claim they are victims of reprisals.

Ontario Nurses' Association (ONA)[13]

The Ontario Nurses' Association stressed the particular problems in the hospital sector, claiming health and safety had only recently become a critical issue, largely as a result of workplace deaths during the province's SARS outbreak in 2003. ONA's recommendations can be grouped under several headings, including increased external enforcement of standards through increased inspection by Ministry of Labour officials, greater use of legal and punitive fines for noncompliance, and increased rights for JHSCs. ONA's brief reflected a view that when budgets are tight, safety receives less priority than it should. "We understand that particularly in health care where costs are soaring and there are innumerable competing interests for money, safety is rarely considered. When it is, the prime motivator is money" (ONA, 2010: 36). ONA was strongly critical of the IRS and JHSCs. They argued, "The internal responsibility system (IRS) is not popular with the labour movement because it is not working. But nowhere is it more dysfunctional (if it even exists) than in the high-risk health care sector" (ONA, 2010: 48).

ONA believes that the "external responsibility system" of enforcement (primarily by MOL inspections, orders and prosecutions), bolstered by other "enforcement"

---

[13] Ontario Nurses' Association, Submission to the Expert Advisory Panel to Review Ontario's Occupational Health and Safety System, June 28, 2010: http://www.ona.org/documents/File/politicalaction/ONASubmission_ExpertAdvisoryPanelToReviewOntarioOccupationalHealthAndSafetySystem_201006.pdf

tools, must be used to stimulate a sluggish IRS and motivate workplace parties to work together to establish safe and healthy workplaces (ONA, 2010: 60).

While critical of the current structure of the IRS, ONA did hold out hope the system could be reformed.

> ONA believes that rather than abandon the IRS, it should be bolstered as a foundation for occupational health and safety "enforcement." There are simply too many workplaces and too few enforcement officers to leave occupational health and safety compliance measures to external parties (ONA, 2010: 61).

Their suggestions to improve the performance of the IRS included paid full-time health and safety representatives at all workplaces and increasing the number of worker safety representatives. JHSC's right to investigate critical injuries needs to be enforced and their rights expanded to include investigation of trends and near misses. Workers should become the majority on JHSCs, and worker representatives should be empowered to issue safety notices directly to employers who are violating health and safety regulations. Certified JHSC members should be given the power to stop dangerous work. JHSCs need to be given more funding from both the employer and safety organizations to perform their work.

Fear of acting on occupational health issues was a major theme in their brief. "We have also received reports of workers who feel intimidated about raising concerns, let alone calling enforcement agencies or exercising their *limited* right to refuse unsafe work. . . . The fear of reprisal is real in our workplaces" (ONA, 2010: 43).

ONA was sensitive to the issue of vulnerable workers. Where vulnerable workers are employed, special attention should be given to the training of workers and educating employers of their responsibilities. While anonymous tip lines can help, in many workplaces these calls are often not really anonymous, hence there was a need for more direct intervention by enforcement agencies. ONA suggested that employers should be discouraged from employing agency workers. Training must be given to agency workers each time they change jobs. To further protect vulnerable workers, ONA felt it was important to include health and safety standards in all supply contracts.

ONA supported an incentive system to encourage safe practices, but saw the WSIB New Experimental Experience Rating (NEER) program based on reported lost-time injuries as flawed. They had strong views on moving prevention activity to the Ministry of Labour and away from the WSIB, which, in their view, was distorting data to make health and safety conditions look more favorable than they were. They envisioned a Ministry of Labour gathering its own data on accident rates, using more enforcement tools to ensure compliance, and working more closely with JHSCs. They viewed administrative penalties (fines for noncompliance) as useful incentives for most health and safety violations, but advocated greater use of prosecution for serious breaches that have or might lead to critical or fatal injuries.

170 / SAFETY OR PROFIT?

Ontario Public Service Employees Union (OPSEU)[14]

The OPSEU brief also called for significant increases in external regulation and changes to the internal responsibility system. Proposals to strengthen external regulation included increasing fines for noncompliance, increasing the number of inspectors and health and safety experts, simplifying the prosecution process, and allowing appropriate cases to go forward under the Criminal Code. The IRS should be strengthened by empowering worker members of JHSCs to stop dangerous work, increasing the obligation of employers to implement recommendations from JHSCs, giving the JHSC a greater role in developing and implementing health and safety policy and improving the training of JHSC members. A key proposal was giving Ministry of Labour inspectors more power to investigate alleged cases of reprisals against workers, the power to reinstate workers claiming a reprisal with pay, and having the Ministry of Labour prosecute reprisal cases.

The OPSEU brief began with a strong condemnation of giving the WSIB the mandate to prevent accidents.

> One approach to clarifying and aligning the roles of the MOL and WSIB would be to move prevention out of the WSIB altogether allowing it to focus on its key business which is administering the province's no-fault workplace insurance system. OPSEU suggests that the Panel seriously consider resituating prevention (including certification standards) to a new branch of the Ministry of Labour. Given the MOL's role to set workplace standards to protect worker health and safety, it makes intuitive sense that the ministry should play an important role in encouraging the improvement of health and safety conditions, and thus prevent injuries and illnesses (OPSEU, 2010: 2).

They were also supportive of a return to greater external enforcement of safety regulations.

> The Ministry's capacity to enforce the *Act*, its regulations and the *Criminal Code* in relation to serious health and safety violations must be enhanced. . . . OPSEU and Labour have consistently argued that the most effective incentive for employers to improve health and safety is a strong enforcement system based on the principle that the cost of violating the law is greater than the cost of compliance. . . . Strong enforcement is vital to address the imbalance of power in the workplace. The IRS is predicated on the erroneous assumption that when dealing with workplace health and safety issues, all the workplace parties are equal. Even in unionized workplaces, workers know that is not true. In many workplaces, unionized or not, workers are afraid to raise health and safety concerns, to demand their rights under the *Act,* and to report workplace injuries and illnesses. With no effective protection against employer reprisals for health and safety activity, workers depend on the enforcement agency for support. In

---

[14] Ontario Public Service Employees Union, Submission to the Ontario Ministry of Labour, June 30, 2010: http://www.opseu.org/hands/pdf/OPSEUExpert%20Panel%20Submission-FINAL-June30-10.pdf

workplaces which make up the large underground economy and those dominated by migrant labour, new Canadians or part-time precarious workers, the need for a strong enforcement system is even greater (OPSEU, 2010: 5).

OPSEU devoted a large section of its brief to the issues of vulnerable workers. Unlike the employer briefs, whose main concern was the cost imposed on them by employers operating in the underground economy, OPSEU was more concerned with the inability of vulnerable workers to make use of the regulatory tools available to them. In their view, the main problem was not lack of knowledge among vulnerable workers, but lack of power to demand improvements. On setting up tip-lines to empower vulnerable workers, OPSEU suggested that this

> is like asking an ant to overpower the giant. . . . Although enabling and protecting workers to inform regulators about underground employers is an admirable goal, other factors in the system need to be fixed before expecting a tip-line to actually help. For example, fear of reprisals from the employer together with a lack of adequate and speedy redress in Ontario to deal with reprisals is a monumental hurdle even if workers are aware of their employer's underground economy status. Indeed, leaving it to workers to carry the burden of reporting on their boss is irresponsible (OPSEU, 2010: 12).

On vulnerable workers the brief indicated

> OPSEU believes that a vigorous, well resourced enforcement system will act as a strong motivator for employers to protect the health and safety of these [vulnerable] workers. OPSEU members, who we consider to be vulnerable workers as described above, would benefit greatly from a more vigorous and enhanced enforcement system, particularly if *OHSA* s.50[15] was enforced aggressively (OPSEU, 2010: 15).

The brief made further suggestions regarding the protection of temporary workers and workers in other nonstandard forms of the employment relationship.

> OPSEU also believes that strong protections are necessary for workers who work for temporary agencies, labour brokers and, those working in subcontracted arrangements. This would include ascribing full legal responsibility to the receiving employer or main contractor for the health and safety of all those working within their sphere of control. For example, the receiving employer for a temporary agency worker would be legally responsible for ensuring that the worker is trained and properly protected from hazards in the workplace. Responsibility for the health and safety of vulnerable workers can no longer be bounced back and forth between two or more agencies, leaving workers completely unprotected and vulnerable (OPSEU, 2010: 15).

---

[15] Section 50 of the act deals with penalties for employers who carry out reprisals against workers who make health and safety demands. One of labor's consistent complaints is that Section 50 is rarely enforced. This was the subject of a major OFL study that was the basis of its brief to the panel. See McCutchen, 2009.

172 / SAFETY OR PROFIT?

As is the case with most unions, OPSEU argued experience rating schemes administered by the WSIB distort the behavior of some employers resulting in claims avoidance, but not necessarily safer working conditions. Rather than monetarily rewarding employers who reduce lost-time injury rates, the OPSEU brief suggested a campaign that would publically recognize good health and safety performance and shame those who continue to operate unsafe workplaces.

> It is possible that publicizing the names of good and bad employers according to compliance with health and safety law may have some utility in achieving compliance. Some firms may want to avoid having their bad behaviour in the public eye. As with environmental standards, society is placing greater emphasis on ethical and socially responsible investment policies. A poor record in this area may deter investment and future business success (OPSEU, 2010: 18).

The OPSEU brief argued that the effectiveness of the IRS was predicated on both strong external enforcement of safety standards and changes to the structure and power of JHSCs.

> The power imbalance between employers and workers needs to be addressed when examining whether JHSCs are an effective mechanism to improve workplace health and safety. Another problem is the growing trend to individualize responsibility for health and safety within the workplace. Rather than the JHSC being considered as the "cornerstone of the IRS," the IRS is increasingly interpreted as a system of individual responsibilities within a workplace. This principle of accountability without power accompanies the downward assignment of responsibility onto workers (OPSEU, 2010: 18).

Other suggestions for JHSCs included a legal obligation for employers to respond to workers and JHSC safety concerns, giving workers a majority on JHSCs and authorizing certified worker reps to issue provisional notices of improvement. In their opinion, external enforcement must drive the IRS.

OPSEU was critical of training policies that limit worker choices regarding how the training is delivered and the trend toward e-learning as an alternative mode of delivery.

> Just as members of professional associations or trades must keep certifications active to effectively contribute to their field; so must JHSC members and health and safety representatives have relevant certification to effectively carry out their functions in the workplace. OPSEU also agrees with the Building Trades that workers should have a "training passport" that lists all the training received, date, trainer, and the delivery organization (OPSEU, 2010: 30).

In conclusion, the OPSEU brief laid out a framework for radical changes to health and safety regulations.

> Every day, OPSEU receives calls from concerned workers who have reached an impasse while struggling to make improvements in their workplaces. Workers efforts are blocked for many reasons: the joint health and safety committee is not

functioning; when health and safety representatives raise concerns, they fall on deaf ears; supervisors are not aware of their obligations under the *Occupational Health and Safety Act;* or, workers are suffering reprisals for exerting their rights under the *Act.* . . . Not only do workers often reach a dead end with their efforts, they frequently do not get the enforcement activity from the Ministry of Labour they need and are entitled to. . . . We believe that now is the time to improve the province's health and safety system. Now is the time to eliminate the culture of fear and truly protect workers from reprisals. Without doing so nothing else will succeed. Now is also the time to strengthen workplace joint health and safety committees and health and safety representatives on one hand and make external enforcement a real possibility in cases of non-compliance on the other. Only by addressing all three of these critical issues will we be able to move the yardstick forward (OPSEU, 2010: 32).

There were a number of common themes in the labor briefs. Their briefs were far more comprehensive than employer briefs and heavily reliant on evidence. The brief from the Ontario Nurses Association was 75 pages long, compared with 7 pages from the BCOHS. The labor briefs also gave more emphasis to the issues of vulnerable workers and the spread of precarious employment.

Not surprising, labor organizations were far more critical of the regulation of health and safety in the province. Relative to the employer briefs, they called for more radical changes, including a substantial return to external regulation as a strategy to deal with the increase in the number of vulnerable workers and the spread of precarious employment. The main problem from the perspective of labor was not lack of knowledge or training but rather the power imbalance at work that put profits before safety. The labor briefs remained supportive of the IRS and JHSCs, although they saw a need for significant changes to the rights of JHSCs and the resources available to them. The labor briefs did not address the problem of overlap and duplication of services that was so predominant in the employer briefs.

The Panel's review (rather than reform) mandate played to the more modest demands of employers to reduce costs while proving to be a significant barrier to labor looking for more substantial reforms to the regulatory process.

THE PANEL REPORT[16]

In early December 2010, the review panel issued its final report. The panel completed its work in less than 9 months. They met as a group eight times in that period. Their work was assisted by eight working groups made up of staff from the Policy Secretariat, the Ministry of Labour, the WSIB, the Health and Safety Associations, and other government ministries. The working group findings were presented to the panel, who then drafted the final report. Seven regional consultation

---

[16] Expert Panel on Occupational Health and Safety, Report and Recommendations to the Minister of Labour, December 2010: http://www.labour.gov.on.ca/english/hs/eap/report/index.php

meetings were held; 250 electronic responses to the questions posed in the consultation paper were received along with over 100 formal written submissions.

The Report's authors suggested their recommendations had the potential to, "set
in place an important cultural shift that could not be achieved through any amount
of regulation" (Expert Advisory Panel, 2010: 2). This and subsequent references
to Expert Advisory Panel all refer to the report cited above in note 16. The focus on
culture is an important indicator that the self-regulatory model promoted by Robens
(1972) and Ham (1976) was to continue. In their view, "Dr. Ham got it right" in
that the government could set standards but could not be in every workplace to
enforce them. The IRS would continue to play a key role in making workplaces
safer, but there was a need to reinforce the three fundamental requirements of a
successful IRS first articulated by Ham: the right to know, the right to participate,
and the right to refuse unsafe work (Expert Advisory Panel, 2010: 6). There was to
be no return to external regulation.

A second general observation is that the Report (Expert Advisory Panel, 2010)
focused predominantly on the issue of workplace injuries and fatalities, most of
which were viewed as associated with activities that involve a high degree of risk.
While there were several references to workplace illness, the growing research
literature on psychosocial risks barely receives a mention. One is led to conclude
that the Christmas Eve accident not only focused the Panel's efforts on the issues of
vulnerable workers and the underground economy, but also the type of workplace
injury, namely, injuries related to violent physical accidents. This is little different
from the focus of the Robens (1972) and Ham (1976) reports.

The third general observation is that the trade-off between profits and safety
was in the minds of Panel members. In the introduction to the report, reference
was made to Canadian employers operating in an "increasingly competitive global
context." This led to the argument that any further regulations must be "cost
effective" and that "evidence-based changes must be made in a manner that helps
achieve measureable improvements in the workplace, which in turn, will improve
competitiveness" (Expert Advisory Panel, 2010: 5). In this context, the Panel noted
that spending more money would further improve prevention and enforcement,
however they believed that their recommendations "can be fully funded within
the current spending allocation" (Expert Advisory Panel, 2010: 6). Much of the
Report focused on how to avoid duplication and better organize the delivery of
health and safety services in the province. In this, they followed the proposals
articulated in the employer briefs reviewed above.

Proposals to Reorganize the Delivery
of Health and Safety Services

The proposal to return the prevention function to the Ministry of Labour is not
new. Prior to 1990, prevention was the responsibility of the Ministry of Labour.
In 1990, an arm's-length agency, the Workplace Health and Safety Agency (WHSA)
was created with responsibility for overseeing the prevention function. This body

was disbanded in 1997 by the new Conservative provincial government, and responsibility for prevention was handed over to the WSIB. The Panel's proposal will largely re-create the situation found in the province prior to 1990. The justification for the move includes greater role clarity, better accountability for investments in prevention activities, and better integration of health and safety services offered to employers and workers. What is unclear from the Report (this and all subsequent references to the Report refer to the Expert Advisory Report) is why an agency within the Ministry of Labour will be any more successful in preventing accidents than the WSIB. One might argue that given the WSIB's direct financial interest in fewer accidents, they would be highly motivated to manage this function effectively. A key issue identified in the Report is that the prevention function is currently provided by multiple organizations, including the WSIB, through workplace audits, financial incentives, and rate rebates for reducing lost-time injury rates; the Ministry of Labour inspectors during site visits and through the orders they issue; and the Health and Safety Associations through their training and education functions. While there are proposals to modify the activities of each of these groups, the reality is that there will still be multiple organizations involved in prevention. The justification for removing prevention from the WSIB is that the Board's insurance function and the challenges posed by the unfunded liability detract from its focus on prevention. One might argue that these two factors would sharpen the Board's incentive to reduce injuries and illness to a minimum.

The Report proposed changes to how inspectors perform their duties. Their existing right to advise employers how to satisfy safety standards is to be reinforced, and they are to be given some role in addressing complaints about alleged employer reprisals. There is some disagreement to what extent these are new powers or simply a reaffirmation of powers inspectors already have. Some suggest that inspectors have reduced their emphasis on advising employers how to meet safety standards for fear of opening themselves to issues of liability should accidents happen after the fact. The six health and safety associations[17] will be overseen by the new prevention organization, but they will continue to deliver services as separate entities. The goal is to minimize duplication between the prevention and enforcement activities of the different organizations involved in health and safety. The objective is a more focused set of prevention activities that are better integrated in the OHS system.

The new prevention organization would be responsible for standardizing training and developing a safety awareness and training strategy to build support for an effective IRS. This is to be complemented by public awareness campaigns, embedding OHS awareness in school and postsecondary curriculum, and mandatory training in

---

[17] The report consistently refers to the six health and safety associations, when it is generally agreed there are only four plus two worker-led organizations: the Occupational Health Clinics for Ontario Workers, and the Workers Health and Safety Center. It seems reasonable to assume that the last two are grouped with the HSAs, despite not truly being HSAs. What this implies for their ability to sustain an independent approach to health and safety services is unclear.

176 / SAFETY OR PROFIT?

high-risk sectors so that workers have greater OHS knowledge before they enter a workplace. This would appear to have implications for the autonomy of labor organizations delivering training to workers.

Measures to Support the Internal
Responsibility System (IRS)

There were seven separate recommendations aimed at strengthening the Internal Responsibility System. The Report suggested general support for the IRS from employers and labor, but concern that without adequate knowledge and given the fear of reprisals, the system was not working to its full potential. One of the problems created by the changing organization of work is that the boundaries between employer and employee are less clear. The proposals included updating guides to OHSA and the role of JHSCs. These updates need to clarify roles between a temp agency and the hiring company. It was proposed that a co-chair of a JHSC could submit a written recommendation to an employer if an issue is not resolved at the JHSC. Employers would have to respond within 21 days. There would be mandatory training for health and safety representatives in small workplaces and mandatory health and safety awareness training for all workers and for supervisors. Mandatory entry training for construction workers should be a priority as well as mandatory fall-protection training for those working at heights.

Combating the Underground Economy and
Better Protection for Vulnerable Workers

Four recommendations focused on the issues raised by the underground economy. The Panel proposed creating a single entity to coordinate a provincewide strategy to deal with the underground economy. Steps should be taken to make it easier for health and safety inspectors to identify who is operating in the underground economy and to take steps to reduce the level of activity. Requiring contractors to post Notices of Project on all construction sites, requiring them to register with the Ministry of Labour, and having them post their WSIB number in a prominent place at a worksite were suggested as ways of minimizing activity in the underground economy. Inspectors should target workplaces and sectors operating in the underground economy for proactive inspections after normal working hours.

A further four recommendations were aimed at improving health and safety conditions for vulnerable workers. The main task of improving conditions for this class of workers was handed off to a special committee to provide advice on matters related to vulnerable workers. This is an indication of the Panel's recognition of the seriousness of this issue, but also the need for new approaches that were beyond the capacity of the Panel to define in such a short time. The Ministry of Labour was encouraged to carry out more proactive inspections and periodic enforcement campaigns in workplaces and sectors where vulnerable workers are concentrated. Health and safety information and training needs to be provided in multiple languages. Some of this information could be distributed through community organizations. It was recommended that the Ministry of Labour develop stronger regulations to control key hazards in farmwork. More of OHSA needs to be applied to farmworkers who are currently exempt from many sections of the Act.

Incentives

There was a lengthy discussion of incentives and their effectiveness in making workplaces safer. This is one area in which the labor movement is unanimous that programs such as NEER, which provide employers rebates for reducing lost-time injury rates, are not leading to safer workplaces but instead lead employers to manage claims and put pressure on workers not to make claims in the first place or to return to work before they are fully recovered from their injury. The Panel did accept labor's view that the existing WSIB incentive programs that tie rebates to lost-time injury rates were not working and were not well correlated with improved health and safety conditions (see Tompa et al., 2007) They recommend that the prevention organization and WSIB revise incentive programs to reduce their reliance on lost-time injury and that the WSIB and the new prevention organization share responsibility for future incentive programs. The panel proposed a number of new initiatives to give employers an incentive to improve health and safety conditions, including an accreditation program to recognize good health and safety practices and incentives to reward employers who encourage better health and safety practices along their supply chains.

Enforcement and Penalties

The main suggestions in this area were to reallocate existing enforcement resources to focus on "tough enforcement for serious and wilful contraventions, as well as compliance assistance where guidance and support for employers help achieve compliance" (Expert Panel, 2010: 43). There seems to be little appetite for increasing the number of inspectors or radically increasing their powers. The balance between internal and external enforcement does not appear to be changing, something that will disappoint most labor groups. The Panel proposed a review of the existing system of tickets and fines for contravention of standards but offered no clear guidance. They also encouraged the prevention organization to explore administrative monetary penalties as an enforcement tool, about which the employer briefs voiced strong reservations.

Improved Protection From Reprisals

Improved protection from employer reprisals was a key theme in the labor briefs reviewed above. Current practice has inspectors investigating workplace health and safety issues while the Ontario Labour Relations Board (OLRB) handles complaints of reprisals. At the OLRB, it can take 8–12 weeks for mediation to begin and up to 6 more months if a hearing is needed. Hearings are held in Toronto, requiring the need for many complainants to travel. The Panel suggested the Ministry

178 / SAFETY OR PROFIT?

of Labour and the OLRB should develop an expedited process to resolve reprisal complaints. The Ministry of Labour would play more of a role in investigating cases of reprisals, including interviews; would pass serious cases on to the OLRB, who would hold an expedited hearing; and might even order interim reinstatement of the worker. It was also suggested that the Ministry of Labour should review its prosecution policy and develop guidance for inspectors to file charges. There was a suggestion that complainants should have access to third-party assistance, such as the Office of the Worker Advisor or Office of the Employer Advisor, to guide them through the process.

Small Businesses

The Panel recognized the unique problems faced by employers and employees in small businesses. As employment in this sector is often less secure than in larger firms, the recommendations here speak to the issue of precarious employment. The Panel proposed the formation of a small business advisory panel of labor and employer members to address the unique issues facing small businesses. They also recommended dedicating resources to help small businesses comply with safety regulations.

Cultivating a Health and Safety Culture

As stated above, the Panel put significant emphasis on creating a health and safety culture as a foundation for safer workplaces. The Panel recommended that the Ministry of Education should work with school boards to expand health and safety content at primary and secondary schools. It was suggested that high school graduation should depend on demonstration of knowledge of occupational health and safety and that postsecondary institutions should incorporate health and safety into their curriculum.

CONCLUSIONS

This chapter used the 2010 Ontario review of its health and safety regulations to explore both the limits of the regulatory systems developed in the 1970s and the scope for change. From the employers' briefs, there was evidence of serious questions regarding Ham's (1976) approach, which empowered labor to be an active participant in matters of workplace health and safety. While accepting that there was some role for worker activism on this issue, it is also clear that groups such as the BCOSH were looking for changes that would rationalize health and safety services by reducing overlap, which meant a smaller role for the independent voice that labor has developed over the last 30 years. The BCOSH went as far as to question the actual benefits of the system that had been put in place and stressed that real gains were the result of employer efforts and investments. On changing labor markets, the employer briefs were concerned that employers in the underground economy were avoiding their fair share of the costs of regulating the system and that policies were needed to better police these employers and if possible, shrink this sector.

The labor briefs offered a far more critical assessment of the effectiveness of the existing regulatory system. Rather than worrying about the cost-effectiveness of service delivery, the labor briefs raised more fundamental questions about whether the system was working. The effectiveness of the IRS was questioned, serious problems in the functioning of JHSCs were raised, and the retreat to individualizing workplace health and safety activism and a frustration with management's unwillingness to accept JHSC and labor recommendations was evident in the briefs. The labor briefs also argued that changes in the structure of the Canadian economy since the 1970s were creating more stress on the system than was apparent in the employer briefs. The general view of labor was that it was unrealistic to expect even a reempowered IRS and JHSC system to effectively shield precarious workers from workplace risks. The preferred solution from labor's viewpoint was an increase in external regulation, including more external inspections and larger financial penalties, as well as criminal prosecutions for employers who fail to protect the health of workers. From labor's viewpoint, the problems with the existing regulatory system were far broader than the issue of employers in the underground economy avoiding their responsibilities.

In many ways, the work of the panel was itself shaped by the "safety or profit" constraint. The panel was appointed in the shadow of a looming provincial election in which the ruling Liberal party was expected to suffer a significant defeat and be replaced by a Conservative party with less enthusiasm for regulation.[18] The buzz surrounding the Panel's activities was that it could not recommend changes that might cost more and that the government certainly was not interested in hearing changes that might lead to a new health and safety regulatory body. Labor was also cautious not to place the Liberals in a bad light, given their concern that this might further strengthen the position of the Conservative alternative. The Panel was given the limited mandate of reviewing (not reforming) health and safety regulations with an eye to improved efficiency. This limited focus was almost guaranteed, given the extremely tight timelines the Panel was given to complete its work.

Both labor and management were successful in getting some of the issues they pushed in their briefs into the Report. Employers were successful in having the Panel focus on costs. The general tone of the brief was one of greater efficiency in delivering health and safety services without the need for additional resources. Labor was successful in getting the Panel to consider seriously the question of employer reprisals, although the proposals fall short of what they would like to see. More importantly, large sections of the Report focus on issues related to changing labor-market structures.

The thrust toward centralized coordination of health and safety services is worrisome regarding its implications for the semi-independent labor organizations that deliver services that provide an alternative view of health and safety issues to

---

[18] This election was held in October of 2011, and to the surprise of many political pundits, the ruling Liberals were returned with a third mandate, albeit a minority mandate.

organizations that have closer ties to employer groups. However, the Panel appears not to have accepted two other components of the employers' views. Both the IRS and the JHSC receive significant support from the Panel recommendations, and several of the suggestions should have a beneficial impact on how these structures work. Better training for JHSC members, better access to information, and a modest increase in JHSC activities all speak to a desire to see JHSCs work more effectively. The willingness to attack the problem of reprisals at work is also a signal of a desire for a more activist labor voice in workplace issues.

Precarity, vulnerability, small businesses, and the underground economy were major themes in the Report. While the recommendations certainly do not signal a return to a regulatory regime based on external enforcement of health and safety standards, there is a general recognition that asking precarious workers to self-regulate in the face of significant power imbalances is also not likely to work. In the end, the Panel handed this issue off to a special committee charged with further investigating how to deal with the issue of vulnerable workers and precarious employment. This must be seen as a first step and the opening of a dialogue on how to best modify the regulatory system to deal with these issues.

The Panel's focus on health and safety culture and more training, with external regulation only in the most egregious cases, will be somewhat disappointing to labor. Many of the recommendations related to training, better flows of information, and how to deal with vulnerable workers are related to the "culture" agenda. The Panel proposed not only increasing training for workers at workplaces and for supervisors and managers, but also reaching out to families, communities, and schools in order to shape beliefs and attitudes. They see this as a way of "preparing the next generation to be good leaders." While there are clearly truisms to what is being said, it remains unclear that the main factor contributing to the continued prevalence of high accident rates is the lack of a safety culture. The brief from the ONA (2010) makes it clear that even when hospital employers understand the need for safer working conditions, tight budgets continue to dictate decisions.

While Robens (1972) saw employer and worker apathy as critical problems in reducing injury rates, and Ham (1976) viewed a need for a stronger labor voice in these matters, the Review Panel (2010) seems to see lack of knowledge as a critical factor. This brings us back to Nichols and Armstrong's (1973) original observation that Robens' apathy argument failed to appreciate the impact of the social relations of production in shaping health and safety outcomes. There is a need for a culture change in Ontario, but I would suggest that it is not the one the Panel has focused on. What is needed is a culture change that shifts the balance between profits and safety and puts greater value on allowing workers to work without shortening their lives. As suggested by Nichols and Armstrong over 3 decades ago, a safety culture will bear fruit only if it can overcome the constraints imposed by competitive markets and the social relations of production.

CHAPTER 9

From *Piper Alpha* to *Deepwater Horizon*

Charles Woolfson

The *Deepwater Horizon* disaster did not have to happen. It was both preventable and foreseeable. That fact alone makes the loss of the eleven lives, the serious injuries to others on the rig, and the ensuing damage and suffering created by the blowout all the more tragic. That it did happen is the result of a shared failure that was years in the making.
—William K. Reilly, co-chair of the National Commission on the BP *Deepwater Horizon* Oil Spill and Offshore Drilling

INTRODUCTION

There is in every disaster an awful sameness that makes comparison a somewhat depressing exercise. In one sense, this is a truism, which at the same time obscures important particularities. Nevertheless, in every case, contingent circumstances combine with a constellation of deeper underlying forces that propel events toward a catastrophic outcome. With hindsight, as every subsequent post-disaster inquiry monotonously reiterates, that outcome was far from inevitable. Yet, as events unfolded, there seemed little that could have halted the spiral to disaster. Whatever the immediate errors of judgement or circumstantial response to unforeseen events, the underlying systemic causes of failure meant that sooner or later, catastrophe was an inevitability. Such a "structuration of failure" requires explanation. It is here that analysis must probe beneath the contingent and immediate causes to examine the contested realms of the political economy of an industry, its relations with the state, the culture of its managerial practices, and its system of its industrial relations, and not least, the formation and deformation of its systems of regulatory oversight. Yet underlying predetermining factors, no matter how perverse their outcomes, must also acknowledge the role of human agency. Individual concrete corporate actors made individual decisions that raise questions of individual and corporate accountability. In the final reckoning, one enduring tension shapes and permeates the

182 / SAFETY OR PROFIT?

ramified chain of circumstance and causation, a conflicted interrogative that is as pertinent today as when the question was first put on the agenda of sociological inquiry: *safety or profit?* (Nichols and Armstrong, 1973). The current chapter addresses this deceptively simple question in the context of the offshore oil industry.

OFFSHORE AS A ZONE OF REGULATORY EXCEPTIONALISM

Beneath the immediate and circumstantial similarities that make for comparisons of episodes of industrial safety failure are more enduring structural determinations. The offshore oil industry is one of the world's most powerful corporate players, composed of giant corporations whose reach and influence extends across the face of the globe. Much of the contemporary history of modern warfare can be written as the story of oil and the struggle between nations for control of this crucial resource (Engdahl, 2004). For the key players in the industry, what matters most is that a business environment is created in which regulatory regimes provide the maximum scope for business discretion and the minimum by way of external accountability. Free enterprise, of which the oil industry is a glorious standard bearer, and hostility to regulation have gone hand in hand since the foundation of the industry (Woolfson and Beck, 2005: 1–14). The power equation that has resulted has meant that oil corporations have been unusually able to shape and mold favorable regulatory regimes to facilitate their objectives, whether in terms of fiscal obligations to national states or in terms of the oversight of safe conduct of their operations for both humans and the environment.

Within this equation of contested national sovereignty on the one hand and the freedom of capital to generate profit in a globalized world economy on the other, the word "offshore" has acquired a special meaning. It has come to resonate with the idea of a "permissioning zone," where the sovereign laws, rules, and regulations either do not apply or do so only selectively. Thus, finance is "offshored" to tax havens where banking scrutiny by national authorities is conveniently lax. Jobs are "offshored" by being exported from traditional production heartlands in developed industrial societies to free trade zones in the developing world, where the standards of pay and working conditions, as well as fundamental trade union rights, are suspended. Perhaps the most pertinent example of "offshoring" is the widely practiced device in the maritime industry of utilizing "flags of convenience" states with lower sea-worthiness and safety standards (Lillie, 2010: 690–693).

Viewed from the perspective of the territoriality of the individual sovereign nation state, the dividing line between what is "onshore" and what is "offshore" is not simply geographically defined. It is closely demarcated by the very forces of capital described above, by "capital's search for a 'spatial-juridical fix'" (Lillie, 2010: 683). Thus, while technically sovereign in a geographical sense, the state cedes its regulatory powers to the demands of capital in favor of minimal regulatory interference. Thus, in the words of Lillie (2010: 683), capital exploits situations in which "territorial sovereignty is little more than a convenient fiction." Capital is therefore able to achieve a "bracketing" of its activities, without challenging the right of the state to "carry on discharging traditional roles *as if nothing had happened*" (emphasis added; Palan, 1998: 627, as cited in Lillie, 2010). The bracketing has as its purpose and itself signifies the power of capital to create and maintain the offshore as a *zone of regulatory exceptionalism*. The problem emerges however when something *does* happen.

In the process of extracting mineral wealth from beneath the seabed, the international oil industry was seemingly able to create and maintain the contiguous territorial outer continental shelf precisely as a zone of regulatory exceptionalism, even though, in law, the minerals beneath the seabed remain within the sovereign power of the nation-state. Comparing the UK North Sea and the Gulf of Mexico at two different points in time, we examine what happens when the state, faced with disaster involving loss of life, is no longer able "to carry on . . . *as if nothing had happened*" (Lillie, 2010). This raises the equally acute question of how regulatory regimes can be successfully reconstructed in the aftermath of regulatory breakdown and loss of legitimacy or of whether and under what conditions it is possible to bring the "offshore" back onshore?

The regulatory environment offshore has throughout remained contested and is itself testament to the economic importance of oil for respective national economies. Both the North Sea and the Gulf of Mexico became provinces for exploration and production of hydrocarbons that were deemed politically safe and conveniently situated close to the mainland industrial societies that they served. The search for oil in the Gulf of Mexico was intensified and in the North Sea was initially driven in the oil-dependent economies of the United States and Britain by the price hike resulting from the Arab petroleum embargo in 1973 and, among other geopolitical factors, the Iranian revolution of 1979. There is not space here to further elaborate the global geopolitics of the international oil industry that led it to invest billions of dollars in exploring and then extracting oil from the outer continental shelves of both Britain and the United States. Suffice to say that these imperatives were both politically and economically urgent. What W. G. Carson, in his 1982 seminal work *The Other Price of Britain's Oil*, called the "political economy of speed," has been dictated by the haste to find secure alternative sources of hydrocarbons that could provide both energy for industry and the consumer and lucrative revenues for national exchequers on a previously undreamed of scale.

Two events, the *Piper Alpha*, which in 1988 was consumed by fire and explosion with the loss of 167 lives, and the *Deepwater Horizon*, which in 2010 suffered catastrophic fire and explosion resulting in the loss of 11 lives and 16 injured as well as massive environmental damage, were signal turning points in the history of the offshore oil industry. Each disaster resulted in a searching public inquiry as national governments struggled to come to terms with the enormity of the tragedy and sought to provide public policy recommendations that would prevent a recurrence. In both cases, we have utilized the official reports into these two events as the main sources of evidence and, in the case of *Piper Alpha*, have revisited earlier work in order to provide a template of analysis with which to approach the most recent disaster (Woolfson et al., 1996).

The report of the UK public inquiry under Lord Cullen QC (1990) into the causes and circumstances of the *Piper Alpha* disaster and of the presidentially appointed U.S. National Commission into the *Deepwater Horizon* disaster (2011) are authoritative, informed, and "objective" critiques of the safety failures surrounding the offshore oil industry. Indeed, the latter is a remarkably uncompromising document, comprehensive in scope and written in a manner accessible to the lay reader, which justifies itself as *the* privileged source cited here. It is both a history of the U.S. offshore industry and an analysis of corporate greed and of regulatory failure. Although not necessarily quoted directly, also consulted were various global and national media sources. Several of these have assembled comprehensive news archives, investigative analyses, feature documentaries, as well as online news, which have provided unparalleled "real-time" access to unfolding events. Furthermore, the major protagonists, such as BP, Halliburton, and industry technical bodies, as well as various U.S. congressional committees of inquiry and regulatory bodies, have held hearings, which are available online, and have issued a slew of interim and final reports. Lastly, a growing and critical academic literature has begun to emerge both in books and journals. Indeed, to date, *Deepwater Horizon* is probably the most studied and best documented disaster in human history. Yet, with rare exceptions, it has been perceived as a quintessentially *American* tragedy, even though it has provoked regulators around the world to review the efficacy of their own regimes. What follows is a first attempt to draw some wider comparisons between two major offshore disasters at different moments in time. If there are common lessons to be learned, then herein lies the answer to our overarching question: Why were the lessons of previous disasters not learned?

THE PARADOXES OF OFFSHORE SAFETY

The *Piper Alpha* oil platform was a production installation in the UK sector of the North Sea off the coast of northeast Scotland operated by Occidental Petroleum. *Deepwater Horizon* was an exploration drilling rig in the Gulf of Mexico owned by Transocean and controlled by the BP oil company. A fundamental difference between the *Piper Alpha* production platform and a drilling unit is that the former gathers hydrocarbons and processes them, whereas the drilling rig seeks never to have hydrocarbons on board except for test and sampling purposes. For example, when circulating oil and gas from the reservoir (and returning them) to test the well for flow, surplus gas will be flared off, but no hydrocarbons will be processed or kept onboard. Both production platforms and drilling rigs are highly complex man-made structures, comprising a number of different functions compressed into what is essentially a small island of ceaseless industrial activity, surrounded by water. Hazard is always present arising from the volatile and flammable nature of the product. If not contained, these hazards can have catastrophic outcomes and therefore, ensuring

FROM *PIPER ALPHA* TO *DEEPWATER HORIZON* / 185

safety and securing extraction and production of hydrocarbons represent a continuous challenge for responsible managements.

The *Deepwater Horizon* disaster of April 2010, the worst offshore disaster in U.S. history, was not the world's worst offshore incident in terms of loss of human life. That dubious distinction belongs to the explosion and fire that destroyed the *Piper Alpha* platform in July 1988. In the years that followed, the recommendations of the exhaustive public inquiry under Lord Cullen were meant to provide a turning point in the management of health and safety in the offshore oil and gas industry. Lord Cullen, an eminent Scottish jurist, was hailed widely for providing the industry with a global safety manual. It would ensure that the lessons of the tragedy were learned and that the dissemination of subsequent recommendations would secure a safe and responsible industry in the future. It did not turn out that way.

This chapter addresses a number of paradoxes. The first, arising from comparison of the two disasters, is that the regulatory regime of safety oversight was, in each case, profoundly flawed. "Regulatory capture" meant that those who should have policed industrial safety in the industry had adopted the assumptions of the industry itself and of governments hungry for lucrative licensing fees and revenues, allowing production imperatives to supplant those of safety. In the trenchant words of William Reilly, co-chair of the presidential National Commission appointed to investigate *Deepwater Horizon*:

> For many years we have a situation which is very close to regulatory capture, they [the regulators] have been driven by revenue generation . . . leasing programs have meant that this is the second largest source of revenue for the federal government after the Internal Revenue Service (BBC World News America, 2011).

In his 1990 review of the UK system of offshore safety control, Lord Cullen identified this conflict of purpose between production and safety. He recommended separation of the regulatory oversight of safety from the functions of energy leasing and encouraging hydrocarbon production. The Department of Energy was duly stripped of its offshore safety monitoring functions, which were handed over to a newly created Offshore Safety Division of the Health and Safety Executive. Yet the regulatory arrangements in the U.S. offshore industry over 2 decades later were in essence exactly those that existed *before* the *Piper Alpha* disaster.

This leads to a second evident paradox in comparing the two disasters. Neither Occidental Petroleum, the oil company operator of *Piper Alpha*, nor BP, the operator of *Deepwater Horizon*, were rogue companies whose practices were far out of line with the prevailing standards in the industry at that time. True, BP's corporate reputation had been seriously tarnished by the Texas City oil refinery explosion in 2005, which killed 15 workers, and by a major pollution scandal in Alaska the following year, as a result of a predicted oil spill from a corroded pipeline (Gillard, Jones, and Rowell, 2005). Occidental Petroleum had also had its share of prior safety lapses, including a fatality on *Piper Alpha*, from which important lessons were not learned. Cost-cutting concerns had led to the postponement of crucial

186 / SAFETY OR PROFIT?

maintenance work (Woolfson et al., 1996: 287–288). Rather than being out of the ordinary, however, both companies were typical representatives of the industry of which they were a part, an industry run by giant multinational organizations aggressively pursuing the bottom line in a ruthless competitive environment.

Each case demonstrates the power of global multinationals to shape and distort a regulatory regime in their favor and to adopt production practices that threatened the life and limb of its workers and inflicted untold damage on the wider community. Within the politically secure province for offshore oil exploration—the Gulf of Mexico—the least safe industrial practices and forms of risk management would seem to have prevailed, not as the exception but as the norm, and they did so seemingly oblivious to the lessons of previous disasters. How did this come about?

THE SPIRAL TO DISASTER

On April 20, 2010, a massive explosion ripped apart the *Deepwater Horizon*, stationed above the Macondo exploration well in Mississippi Canyon, Block 252, located on the U.S. outer continental shelf 49 miles off the coast of Louisiana. A total of 11 offshore workers died in the most horrendous of circumstances (National Commission, 2011: i). The rig was leased by the client operator, BP, together with co-venturers, Anadarko Petroleum and MOEX USA. It was managed by drilling contractor Transocean, itself a global player in the industry, which owned and crewed the rig. The global construction contractor, Halliburton, was responsible for key aspects of the drilling operations, in particular, the provision of the vital "cement" to ensure the well's integrity.

At a cost of $350 million, *Deepwater Horizon* was one of the world's most sophisticated and expensive drilling rigs and the pride of the Transocean fleet. It was drilling the deepest offshore well in the history of the industry, using cutting-edge technology to penetrate a high pressure reservoir of oil and gas 5,000 feet below the surface of the ocean and a further 13,000 feet below the seabed. That day in April, the rig completed the final operation, the phase of "well completion," of a successful, if technically very challenging, drilling job. This involved securing the integrity of the well by a cementing process, disconnecting the drill pipe and moving the rig away to allow a production facility at some point in the near future to be located at the site. It had been originally estimated that drilling operations would require 51 days in total for which a budget of $96.2 million had been allocated by BP. It did not turn out that way. The Macondo was "the well from hell," which had presented several costly delays. As a result, there was substantial cost overrun of $58 million on the project and a schedule delay of 6 weeks, costing roughly $1 million per day. What happened next was the inevitable result of the concatenation of pressures emanating in the first instance from the corporate headquarters of BP in Houston, Texas.

One of the final tasks was the completion of "temporary abandonment," moving the drilling rig from over the well head and sealing the bore hole in preparation for a production installation taking over at some future point. Preparing the well for temporary abandonment required a process of cementing in place the pipe casing that ran into the oil reservoir, preventing any oil and gas from flowing up the drill hole. A 3-man crew from the oil industry services company Schlumberger was due to fly out to the rig that same day to perform the tests to ensure that the cement plug 3,000 feet below the top of the well was sound enough to contain the high pressure reservoir of volatile hydrocarbons beneath. When they arrived, they were told that the tests were not required and that the cementing had gone smoothly.

At 11 a.m. that morning, the Schlumberger team were flown back to shore, saving BP both time and a $128,000 fee. This was only the last of a series of safety shortcuts taken under BP's instruction, which were identified by the National Commission in its 2011 report, which included an unorthodox system for fixing sections of drilling pipe in place and alterations to operational procedures that were not or only inadequately approved by the regulatory body, the Minerals Management Service (MMS). Later that same morning, a high-profile VIP group of executives from Transocean and BP arrived by helicopter from Houston to participate in an "management visibility tour," which, among other matters, would celebrate the rig's safety record of 7 years of drilling without a single "lost-time incident." The crew had even recorded a "rap" song to emphasize the safety message. Whatever the truth to the claim of "zero incidents" might have been, the visit by senior management was, in the view of at least one informed industry observer, a "lost opportunity" to identify serious problems of process safety on the rig and one of the last chances to avert the impending disaster (Hopkins, 2011).

Meanwhile, other final tests were carried out by the onboard personnel that day, including so-called positive and negative pressure tests to check the integrity of the well and its metal casing. If the results from either were not as predicted, then there would be obvious grounds for concern. While the positive pressure tests provided unambiguous results, the results from the negative tests were more uncertain. It was expected that during these second tests, the pressure in the well would remain constant, whereas, in fact, it repeatedly built back up, seemingly sufficiently so for there to be at least grounds for concern among some of the drill crew (National Commission, 2011: 5–6). A second additional pressure test was conducted that evening, which again yielded an anomalous reading, but this result was explained by reference to a supposed "bladder effect." Acting on the assumption that all was in order, the crew began final operations prior to disconnect, pumping seawater down the drill pipe to displace the drilling fluid ("mud"), which is essential to lubricating the drill bit along with the capping layer of liquid "spacer" used to separate the seawater from the drilling fluid. Drilling mud changes as dictated by the job it has to do. The main function is to provide hydrostatic pressure in the wellbore. In the early phase of drilling, it must also be capable of lifting drill cuttings to the surface and lubricating and cooling the bit. However, at the time of this blowout, the configuration of the mud was such that hydrostatic balance was its only function. The decision was made by BP personnel to instruct Halliburton to remove heavy drilling mud and replace it with much lighter seawater in order to save time and money. The result however was that the pressure of the column bearing down upon the wellbore no longer exceeded the upward pressure of the build up of gas in the undersea formation.

Sometime after 9 p.m., as the VIP delegation, having dined, now embarked on a tour of the impressive control room on the bridge, a first "high-frequency vibration" was felt, followed by an ominous hissing noise (National Commission, 2011: 8). On the drill floor, mud and seawater gushed up from below. As drilling fluid spewed onto the drill floor, it was realized that a "kick" was in progress, with hydrocarbons quickly spewing upward inside and over the top of the 20-story drilling derrick. The blowout had begun. Within seconds, a massive explosion rocked the rig as escaping gas, which now enveloped the entire rig, found a source of ignition, and the first fire began on the starboard side of the drilling derrick, thereafter enveloping the derrick "in a firestorm of flames" (National Commission, 2011: 9, 13). Gas alarms could be heard as the rig's engines began revving.

The chain of simultaneous and subsequent events quickly took the initial blowout to catastrophic dimensions. The rig was plunged into darkness and a momentary eerie silence as power systems failed before the second in a series of explosions occurred, ripping through the engine control room and destroying large parts of the rig infrastructure. Those who could do so rushed in panic to emergency stations for lifeboat evacuation, taking with them the injured who could be carried, as a general alarm sounded out. From the bridge, a *mayday* message was sent out over the rig radio. What was already clear as the conflagration continued was that a crucial piece of safety equipment, the blowout preventer (BOP), designed to automatically shut off the rig from any uncontrolled flow of hydrocarbons into the well beneath, had failed. The fire had an endless reservoir upon which to feed itself. Yet, even faced with this unfolding catastrophe, the last line of defense, the emergency disconnect system (EDS) to the blowout preventer, was not activated by the crew, who waited for authorization from the rig manager. In a scramble of commands, this authorization was finally given (National Commission, 2011: 13). Assuming that the BOP was now unlatched, some crew members tried in vain to restart the standby generator, but to no avail. The rig still remained umbilically connected to the hydrocarbon reservoir, stoking the exploding furnace that *Deepwater Horizon* had become.

The actual evacuation was chaotic, with lifeboats launched, sometimes only half filled. Some individuals decided to take their chances and jump into the sea. By now, lifeboats were either being launched, with their frightened occupants screaming to go, or they were already in the water as evacuation became the only possible line of action. Yet, even here, tragedy and farce rubbed uncomfortably against each other. The remaining inflatable life raft snagged its rope during launch, and no knife could be found to free it (pocket knives were not allowed on the rig), until someone discovered a suitable mechanical cutting tool. As the inflated raft was pulled away from the disintegrating rig by some of the survivors, now swimmers in the water (no one could find the paddles), the rig's captain, having jumped from 100 feet above, splashed into the water a few feet away. He was to be followed by others of the crew making their last desperate escape. However, even then, escape was nearly doomed as a painter line continued to tether the life raft to the rig. The captain swam to a fast rescue craft launched from the emergency standby vessel to retrieve a folding knife. As the survivors gathered in the galley of the standby vessel, now stationed next to the blazing rig, the headcount began. Of the 126 persons onboard, 11 remained unaccounted for, mainly workers on the drill floor, and 16 were seriously injured. Over the next hours, the disintegrating rig listed heavily and was consumed in a final conflagration that eventually left only a smoking black plume billowing upward from the otherwise calm waters of the ocean into a clear blue sky.

While every disaster has its own unique anatomy of failure, what is striking in comparing the unfolding circumstances of these two offshore disaster events, *Piper Alpha* and *Deepwater Horizon*, is the number of similarities. In the first place, the spiral of events was accelerated by a series of crucial human errors and hesitations at critical junctures. In the case of *Piper Alpha*, the most significant of these was the failure of management on adjoining rigs, faced with the obvious conflagration on *Piper Alpha*, to take the initiative to shut down linked pipeline processes unless authorized by onshore management. As a result, the supply of hydrocarbons from the connected neighboring platform pipelines effectively fed the fire on *Piper Alpha* (Woolfson et al., 1996: 238–239). Moreover, it was clear that those who had authority in both disasters were unprepared for an emergency of this scale. In the case of *Piper Alpha*, there was a failure to provide the crew with instructions to facilitate orderly evacuation, while previous safety instructions proved not to be appropriate or relevant on the night of the disaster. On *Deepwater Horizon*, the bridge did not immediately sound a general alarm to begin evacuation. Gas alarms on the rig were activated as the flammable gas escaped, but as on *Piper Alpha*, the emergency systems that might have prevented the gas from spreading or igniting were not operative. The engine control room on *Deepwater Horizon* was told of a well-control situation, but not of the scale of the escaping gas and mud. When the control room did become aware of the escaping gas, the crew did not shut it down, again awaiting instructions from the bridge. Delay was disastrous as thereafter, the engine control room was the site of two subsequent explosions (National Commission, 2011: 240, 243). Such failures of coordinated emergency response and critical hesitations were the result of a lack of emergency preparedness, disabled individual responsibility, and the fear of making decisions without authorization.

Yet other disturbing similarities mark both events and relate to more underlying problems that are located in the labor relations regime offshore. On both installations, individual operatives had voiced safety concerns over a period of time, which management did not adequately respond to and identify as warning signals that safety systems were deficient. On *Piper Alpha*, the permit-to-work system, which governed the sequencing of hazardous work, had largely ceased to function as an effective means of safe task organization and had already led to one fatality. On *Deepwater Horizon*, members of the workforce had suggested that the safety critical blowout preventer on the rig might have had serious defects, but there was no adequate managerial action to follow. Everything pointed to haste, corner cutting, cost saving, and sloppy management, an underlying commonality linking both *Piper Alpha* and *Deepwater Horizon*.

190 / SAFETY OR PROFIT?

REGULATORY CAPTURE: OFFSHORE UK

The offshore regulatory regime was created as a zone of regulatory exceptionalism in *both* the North Sea and the Gulf of Mexico. This in itself provides a compelling story of corporate power and influence over successive governments on both sides of the Atlantic and of continuing resistance by the industry to more rigorous regulatory controls. For reasons of space here, only the outlines of this saga can be summarized.

The starting point is the notion of regulatory capture, the mechanism with which regulatory exceptionalism was obtained, whereby the agency charged with oversight of safety comes instead to adopt the views of industry and of government itself in the prioritization of production over other concerns (Woolfson et al., 1996: 295–296). This contradiction of purposes between safety oversight and facilitation of production (and revenues) was recognized in the fundamental reconstruction of onshore safety regimes in the UK in the 1970s, but not offshore. The question is how this "anomalous" safety regime offshore came about.

In the North Sea, from the mid-1960s until the defining moment of the *Piper Alpha* disaster over 2 decades later, the regulatory regime was largely reactive and only slowly came to embrace explicit concerns for safety (Woolfson et al., 1996: 249–275). The loss of the *Sea Gem* jack-up exploration rig in 1965 eventually led to the passing of the Mineral Workings Act in 1971, with powers to regulate "the health, safety and welfare of persons on offshore installations." Even then, the regulatory system was largely prescriptive in nature, based on detailed regulation that was often difficult to enforce. Government was keen to reassure the industry that it could rely upon "benevolent enforcement . . . generally advisory in nature" resulting in a regulatory setup that Carson (1982: 152, 231) famously described as the "institutionalised tolerance of non-compliance." This regime was presided over by the Department of Energy, the very same ministry responsible for successive licensing rounds and for ensuring the tax returns to the UK treasury.

Meanwhile, the onshore safety regime, while far from perfect, evolved in a quite different direction away from superficial or tick-box compliance with prescriptive regulation and toward more system-oriented goal-setting regulation as embodied in the Health and Safety at Work Act 1974, which followed on from the Robens Report of 1972. The Health and Safety Commission (HSC) and the Health and Safety Executive (HSE) were created as new independent agencies free from sponsoring control by individual ministries of government. What is striking, however, is that the onshore regulatory regime did not follow the oil industry offshore. By the time the first offshore production installations were pumping oil out of the seabed and exporting it to the UK mainland, two entirely divergent safety regimes existed onshore and offshore, governed by two conflicting styles and philosophies of safety and risk management. Thereafter, there was trenchant industry resistance to successive attempts by the HSC to extend the authority of the Health and Safety at Work Act offshore, despite worrying levels of and a public inquiry into fatalities, particularly in the drilling industry (Woolfson et al., 1996: 267–270). Eventually, in

September 1977, the Health and Safety at Work Act was extended offshore, but on the basis of provision-by-provision approval of specific regulations, while oversight of safety on installations was handed over to the Petroleum Engineering Division, a branch of the Department of Energy under an "agency agreement" with the onshore HSE, thus neatly short-circuiting the possibility of more effective regulatory scrutiny and more advanced safety thinking. Indeed, thereafter, the Petroleum Engineering Division routinely resisted any extension to offshore of onshore safety regulations, which mandated goal-setting systematic approaches to controlling major hazards. A measure of their success in so doing was that even when offshore safety did become a matter of public inquiry in the early 1980s due to the extraordinarily high rate of fatalities among offshore divers, the Department of Energy, with enthusiastic support from the industry, was able to successfully marginalize the HSE and retain the status quo (Woolfson et al., 1996: 260–273).

In developing their congenial relationships with the regulators, the oil companies also successfully aimed to convince government that the Department of Energy possessed "special expertise" and was therefore the appropriate organization to regulate offshore, since it had, in the words of Shell Oil, "grown up with the offshore industry" (Woolfson et al., 1996: 264). Ensurance of remaining compliance of regulators with industry needs was further assisted by the direct recruitment of top officials from the Department of Energy by the industry in a hiring system perhaps best described as "deferred bribery." Offshore inspections were not surprise visits but relied on the provision of transport by the oil companies for inspectors wishing to go offshore. Due to shortages of manpower, serious accidents and even fatalities were investigated for the lessons that could have been learned in less than half the possible cases. As it happened, *Piper Alpha* had an inspection visit a mere 11 days before the disaster, an inspection subsequently described by Lord Cullen (1990: 48) as being "superficial to the point of being of little use as a test of safety on the platform." Meanwhile, the Petroleum Engineering Division, although grossly understaffed, refused temporary placements from the Health and Safety Executive and, at the time of *Piper Alpha*, had a complement of only five inspectors to cover 139 fixed installations and 76 mobile rigs on the UK continental shelf. Undermanning not only affected the frequency but the depth of inspections, leading Lord Cullen (1990: 48) to observe that "in my view the inspectors were and are inadequately trained, guided and led."

There were also clear deficiencies in the system of permissioning overseen by the Department of Energy. Carson (1982: 242) documented that permissions were granted to operators to use "temporary accommodation" modules, which had inadequate fire and blast protection long after the rig construction phase offshore had ceased and the more hazardous drilling and production operations had commenced. The companies themselves sought successive extension of permissions to use temporary accommodation modules even in the face of individual Department of Energy inspectors' concerns over safety (Carson, 1982: 244). It was in the accommodation module on *Piper Alpha*, which proved woefully inadequate to withstand the fires on the rig, that many of the crew died. Yet the accommodation

192 / SAFETY OR PROFIT?

module, according to safety instructions, was supposed to serve as the muster and control point for evacuation from the rig. Doubts about its fire protection capabilities had been raised with the operator by inspectors from the Department of Energy as far back as 1975, but to no effect, as further exemptions were granted for the next 13 years, overruling expressed concerns of Lloyds' Register, the external certifying authority. The disastrous consequences of the deficiencies of this regime were forensically exposed in the public inquiry under Lord Cullen in the aftermath of the *Piper Alpha* disaster (1990: 20). The oil operators had successfully created and sustained a regulatory regime with outdated prescriptive approaches to safety. It remained so largely unhindered by external regulatory interference until *Piper Alpha* exploded some 13 years later.

REGULATORY CAPTURE: THE GULF OF MEXICO

The Gulf of Mexico has long been a site of oil production in the United States. In the 1990s, technological advances in underwater seismic imaging and drilling enabled the opening up of new oil reserves trapped beneath massive sheets of salt deposits at depths of up to 10,000 feet below the surface of the sea in the "ultra-deepwater" of the Gulf. Such was the success in overcoming the immense technical challenges posed by deepwater drilling, that by the end of the decade, deepwater overtook the volumes of shallow water oil extraction (National Commission, 2011: 41). The company leading the way in opening up these new reserves was BP, now excluded from previous areas of high profit such as the Middle East and Nigeria, and with declining production in the North Sea.

The Outer Continental Shelf Lands Act of 1953 placed responsibility for the development of offshore mineral extraction under U.S. sovereign jurisdiction with the Department of the Interior. While local states resisted federal encroachment, Supreme Court rulings eventually gave the federal government control of the outer continental shelf area beyond a 3-mile offshore limit. When the first outer shelf area leases were offered for sale, they were to prove immediately highly lucrative for the federal government. Leasing policy thereafter remained largely unchanged until the end of the 1960s, when a blowout offshore in California and the rise of the environmental movement in the United States combined to produce the National Environmental Policy Act (NEPA) of 1970, the first and most important of many such environmental protection laws of the decade (National Commission, 2011: 57). It was only after the oil embargo of 1973 had given specific and urgent impetus for the drive for energy self-sufficiency that the Outer Continental Shelf Act of 1987 provided the Department of the Interior with the impetus and authority to expand offshore drilling (National Commission, 2011: 60–61).

With the arrival of the Reagan administration in the first years of the 1980s, a massive and lucrative expansion in offshore leasing took place. The Minerals Management Service, (MMS) was the main regulatory federal agency responsible for overseeing this offshore province. The MMS originated a context driven by that administration's desire to decrease reliance on foreign energy supplies and at the same time, ensure the financial fruits of its plan for a massive expansion in offshore drilling. New arrangements were put in place known as "area-wide leasing" (AWL), opening up "larger areas of choice to industry . . . shifting environmental and resource assessment to the post-lease phase. The logic of AWL was to 'explore first and ask questions later'" (Priest, 2010). Thus, under the authority of the Outer Continental Shelf Act, the MMS came into being in order to facilitate the more efficient gathering of revenues for government from the accelerated procedures of offshore leasing concessions.

An inherent contradiction between stronger regulation and accelerated development was built into the program for developing the offshore continental shelf from its inception in the 1970s and "has bedevilled it ever since" (Priest, 2010). In the words of the National Commission (2011: 56), "environmental protections and safety oversight were formally relaxed or informally diminished so as to render them ineffective." With the dramatic increase in oil prices over the previous decade, by the 1980s, royalties and revenues from federal oil and gas resources had already become the second largest revenue source for the U.S. Treasury. The result was that the MMS became responsible for regulatory oversight of safety in offshore drilling and for collecting revenue from that drilling in which the latter became "the dominant objective" (National Commission, 2011: 56).

The subsequent move into deepwater drilling brought with it not only increased technical challenges but increased risks, known and widely discussed within the industry. However, increased risks were not matched by intensified regulatory oversight. Indeed, the reverse was the norm in an eerie echo of how the industry majors had responded with sustained opposition to the possibility of greater regulatory oversight in the UK offshore sector in the years prior to *Piper Alpha*. The industry, in the National Commission's words (2011: 56), "regularly and intensely resisted such oversight, and neither Congress nor any of a series of presidential administrations mustered the political support necessary to overcome that opposition." Nor did the industry itself, despite assurances to the contrary, make significant investments in drilling safety and oil-spill containment technologies. In an indictment as damning as that which Lord Cullen delivered with respect to the failures of the UK Department of Energy, the National Commission (2011: 57) concluded that "for a regulatory agency to fall so short of its essential safety mission is inexcusable."

Here then was a classic iteration of regulatory capture. The agency responsible for policing the industry had come to adopt the assumptions of the target industry as defining and coterminous with the public good. So blatantly compromised was the regulator that President Obama was moved to comment on the "too cosy relationship" between the industry and the MMS. More than shared assumptions and the subversion of priorities, it also in some instances involved payments of royalty revenues "in kind" rather than in cash, shared hospitality, and other favors exchanged between the industry and regulatory personnel, including relationships of industry personnel with the "chicks" of the MMS (Christian Science Monitor, 2010; National Commission, 2011: 77–78). How had this state of affairs come about?

194 / SAFETY OR PROFIT?

Even in terms of oil exploration on the U.S. continental shelf, the Gulf of Mexico was a zone of regulatory exceptionalism. The southern coast of the United States produces one third of all domestic oil, and in the rush for "energy security," the Gulf was an area of exemption ("categorical exclusion") from the more onerous environmental regulations that had inhibited offshore developments elsewhere in the United States. Under the Clinton administration, categorical exclusions granted in the central and western Gulf rose from 3 in 1997 to 795 in 2000. During the following Bush years, the MMS granted an average of 650 categorical exclusions per year in the region, falling to 220 during the Obama administration's first year (Eilperin, 2010). Offsetting the apparent imperative of environmental review therefore, a "carefully calibrated political compromise" was designed to promote offshore drilling in the Gulf by effectively creating tacit exemption from the NEPA requirements for prior environmental impact assessment (National Commission, 2011: 82). By so doing, the delay between exploration and production of 3 to 6 years was significantly reduced and "burdensome" and "unnecessary delays for operators" were avoided (National Commission, 2011: 62, 82). This exemption, set in place in the early days of deepwater exploration, also applied in the specific case of BP's exploratory Macondo well concession, whereby the MMS "categorically excluded from any NEPA review the multiple applications for drilling permits and modifications of drilling permits associated with the Macondo well" (National Commission, 2011: 83). Behind this effective waiver of statutory requirements of environmental impact assessments was the rationale that the Gulf was a mature area of oil activity in which the risks were already well known in comparison with "frontier areas."

Thus it was that regulatory oversight of the industry and the collection of revenues from that industry were combined in the same agency, the MMS, effectively replicating the inherent contradictions that beset the UK Department of Energy in its offshore role. However, the federal authorities were not to have it all their own way, and an increasingly hostile Congress imposed moratoriums on the Department of the Interior's budget, with the practical effect that expansion of offshore drilling was inhibited, and accordingly, the Gulf of Mexico assumed a "still-more-special status" (National Commission, 2011: 66). It became one of the few areas within U.S. jurisdiction not subject to prohibitions on new leasing activities and exploration and development of existing leases. The zone of regulatory exceptionalism *par excellence* therefore was the Gulf of Mexico.

THE MMS AND OVERSIGHT OF
OFFSHORE SAFETY

How practically did the MMS perform its contradictory role under these circumstances? The National Commission (2011: 67) was clear that the root problem was not the lack of governmental sovereign authority as such—the question of who owned the nation's natural resources was solved—but rather that:

political leaders within both the Executive Branch and Congress have failed to ensure that agency regulators have had the resources necessary to exercise that authority, including personnel and technical expertise, and, no less important, the political autonomy needed to overcome the powerful commercial interests that have opposed more stringent safety regulation.

Such regulations as existed were highly prescriptive, referring to detailed specifications and technical requirements for pollution prevention and control, well-completion operations, major maintenance, production safety systems, platforms and structures, pipelines, well production, and well-control and production safety training. Even so, unwelcome "prescriptiveness" was easily circumvented. Take for example the regulations governing the critical blowout-preventer equipment. The industry contended that its own standards were more reliable than the regulations and thus required less frequent pressure testing (a process that would interrupt other activities). The MMS acceded to the industry view and reduced the number of mandated tests by half. When third-party technical studies pointed to possible failures in blowout-preventer systems, the MMS commissioned its own studies, which found that many rig operators were not testing BOPs and were basing claims of safety "on information not necessarily consistent with the equipment in use" (National Commission, 2011: 74). Transocean, the owner of the *Deepwater Horizon* rig, failed to recertify the BOP for over 10 years, despite recommendations from the manufacturer and industrial bodies for recertification at 3- to 5-year intervals. Important BOP parts were not replaced according to recommendations. Yet the MMS never revised the regulations or instituted independent inspection and verification procedures. Again, there were no meaningful regulations governing the requirements for well cementing and testing. Nor were there regulations governing negative-pressure testing of well integrity—a fundamental check against dangerous hydrocarbon incursions into an underbalanced well. Instead of setting the parameters for offshore safety, the MMS was dependent on "industry standards" devised by the industry itself. American Petroleum Institute records show that the MMS adopted at least 78 such industry-generated standards as federal regulations (Eilperin, 2010). On many safety-critical matters, it is unsurprising that the federal regulations "either failed to account for the particular challenges of deepwater drilling or were silent altogether" (National Commission, 2011: 228).

Moreover, the agency's resources did not allow it to keep pace with the expansion into the deeper waters of the outer continental shelf, as "senior agency officials' focus on safety gave way to efforts to maximize revenue from leasing and production" in what the National Commission (2011: 68, 76) called "a culture of revenue maximization." Indeed, just as the industry moved into deeper waters and technological developments accelerated, the MMS and its junior regulatory partner, the U.S. Coast Guard, suffered severe budgetary and staffing constraints (and in the case of the latter agency following the September 11 attacks in 2001, a reorientation of priorities to homeland security). These constraints on resources resulted in a lower level of oversight with one inspector for every 54 offshore facilities in the Gulf, compared with the Pacific Region employing five inspectors to inspect 23 production facilities. In many cases, a lack of technical training for the inspectorate in critical aspects of rapidly advancing drilling technology made effective inspection difficult, with inspectors relying on oil company expertise and information. The already small number of unannounced visits declined further, and scrutiny of permit applications by the oil companies, according to standardized and consistent procedures, was increasingly problematical to sustain. Much of this was familiar in the UK sector of the North Sea, as previously detailed. However, even more fundamental was the question of the safety philosophy underlying the respective inspection regimes, and here again, the UK experience presaged much of what was still the prevailing approach on the U.S. continental shelf. In the words of one MMS official in 1996, later cited in the National Commission (2011: 71–72), "We want to approach our relationship with the offshore industry more as a partner than a policeman."

In his critique of the UK offshore regulatory regime under the Department of Energy, Lord Cullen's 1990 report into *Piper Alpha* had specifically identified the failure of the regulator to adopt a modern goal-setting safety management approach, relying instead on an outmoded tick-box approach and detailed prescriptive regulation. Little or nothing, said Lord Cullen (ch. 22.21) had been learned from the more developed onshore goal-setting approach of the Health and Safety Executive based on a holistic approach to safety management, or from the contiguous offshore Norwegian regulatory system, with its systematic approach to risk management based on "internal control." His remarks on the failures of prescriptive regulation could have applied with equal force to the extant U.S. offshore regulatory regime 20 years later.

Lord Cullen's key recommendation (1990: ch. 17.37) was a "safety case" approach for the UK offshore regime that would allow comprehensive and effective risk management throughout the life cycle of an offshore installation and demonstrate that the operator had considered and documented by systematic risk assessment the various hazards facing their offshore installations and personnel. Underpinning the safety-case regime were new sets of specific regulations designed to give the new regime "solidity" dealing with fire and explosion protection, evacuation, escape and rescue, and other key safety issues. The whole reconstructed safety regime, which Lord Cullen saw as not taking place overnight, required the approval of a new independent regulator, the Offshore Safety Division of the Health and Safety Executive.

In the industry globally, formal safety assessment in the shape of a safety-case regime was already conventional wisdom following the *Seveso* disaster in Italy in 1976, and since then enshrined in the safety management protocols of high-hazard onshore petrochemical establishments. However offshore, it was an approach that was flagrantly ignored, particularly in some of the murkier, more compliant Third World regulatory regimes in which the international oil industry operated. That the U.S. offshore authorities and those same offshore operators had seemingly remained "impervious" to modern safety thinking in the offshore First World speaks

FROM *PIPER ALPHA* TO *DEEPWATER HORIZON* / 197

volumes to the power of the multinationals to create a selective exclusionary "regime of exceptionalism."

Even tentative steps toward requiring operators to produce a safety and environmental management program (SEMP), which would have acknowledged some of the reforms introduced offshore elsewhere after major disasters such as *Piper Alpha*, remained, as the National Commission (2011: 71) put it, "indefinitely frozen in time." There were extended renewals in the lengthy consultation process with the industry as the regulators advanced the possibility of a new approach to safety and environmental management, leading the MMS to urge companies to adopt such systems voluntarily and thereby prevent formal regulatory intervention. But even this was not enough. When the MMS, in 2003, proposed to update its requirements for the reporting of key risk indicators (all unintentional gas releases to be reported, because even small gas leaks can lead to explosions), the White House stiffly opposed (National Commission, 2011: 72).

The U.S. offshore industry, with the consent of successive federal governments, was itself therefore entirely complicit in resisting efforts to reform, which might have encroached upon its freedom to operate in the manner it thought best. The threat of regulatory reform being imposed was effectively torpedoed by industry promises of self-regulation through the creation of voluntary guidance or industry-led performance standards. In any case, such voluntary guidance either failed to emerge or did so in such a tortuously slow manner as to render the status quo largely unaffected. In this respect, the U.S. offshore industry exactly paralleled the pattern of studied resistance to regulatory renewal that characterized the UK offshore sector prior to *Piper Alpha*. The American Petroleum Institute (API), the International Association of Drilling Contractors (IADC), and national associations of major oil operators, operated as alert in-house watchdogs for the industry, guarding the regulatory boundaries of the offshore zone of exceptionalism from unwelcome regulatory encroachment. The National Commission (2011: 228) observed,

> Beginning early in the last decade, the trade organization (API) steadfastly resisted MMS's efforts to require all companies to demonstrate that they have a complete safety and environmental management system in addition to meeting more traditional, prescriptive regulations—despite the fact that this is the direction taken in other countries in response to the *Piper Alpha* rig explosion in the late 1980s.

Until the Macondo well blowout consumed *Deepwater Horizon*, the U.S. regulators' efforts to devise a more effective safety regime had repeatedly failed. At the time of the blowout, MMS had not published a rule mandating that all oil rig operators have detailed plans to manage safety and environmental risks—more than 20 years after a rule was first proposed. In the words of the National Commission (2011: 71), such proposals "were repeatedly revisited, refined, delayed, and blocked alternatively by industry or skeptical agency political appointees."

198 / SAFETY OR PROFIT?

POST-DISASTER REGULATORY RECONSTRUCTION
OF THE OFFSHORE REGIME

Faced with the comprehensive failure of the existing regime of regulatory oversight, industrial disasters on the scale of *Piper Alpha* or *Deepwater Horizon* call forth demands for public policy interventions of a far-reaching nature. Piecemeal reforms are seen as inadequate, and "a return to the drawing board" is often mandated. In the case of *Deepwater Horizon*, protracted attempts to kill the well were relayed on nightly television for 152 days, until a relief well begun in early May was ultimately successfully drilled. News reports depicted crude oil escaping uncontrollably from the seabed, and gooey slicks seeping ashore on the pristine resort beaches and encroaching on the sensitive environmental wetlands of Louisiana and Florida ignited a national fury against BP. The seemingly futile efforts at containment using shoreline booms, airborne chemical dispersants, and oil skimming from the surface of the sea suggested a patently inadequate oil spill response plan in which local, state, and federal officials struggled to coordinate their efforts. The sight of oil-sodden seabirds gasping for survival along the shorelines summed up BP's loss of corporate credibility more than anything. The company stumbled to manage its response under the inept leadership of its harassed chief executive, Tony Hayward, who was quoted as saying, "There's no one who wants this over more than I do. I would like my life back" (BBC, 2010). BP appeared incapable of containing the oil still gushing out from the well head as initial estimates of the "modest" scale of the spill were repeatedly revised upward from the hundreds to thousands and ultimately tens of thousands of barrels of oil a day (National Commission, 2011: 147).

As the days turned to weeks, and with no conceivable end in sight, and repeated unsuccessful attempts to staunch the flow of oil into the Gulf, on May 2nd, the president himself embarked on a hurried catch-up fact-finding tour to the Gulf coast, the first of several. This brought him face-to-face with the reality of impoverished shrimp boat fishermen from devastated local coastal communities. President Obama quickly echoed the nation's popular outrage in famously unpresidential words: "I want to know whose ass to kick" (Guardian, 2010). Such refreshing calls for accountability are of course always welcome from the Executive Branch of government. Unfortunately, the answer seemed to lie, at least in part, uncomfortably close to home. In the words of the National Commission, the disaster was created not only by corporate mismanagement, but by "*failures of government to provide effective regulatory oversight of offshore drilling*" (2011: 122; emphasis added).

In the introductory section of this chapter, the question was posed as to what happens once the state is no longer able "to carry on discharging traditional roles *as if nothing had happened,*" when sovereignty can no longer remain simply a "convenient fiction" and its absence becomes an "inconvenient" fact that must be somehow contemplated. A disaster on the scale of *Deepwater Horizon* makes carrying on "as if nothing had happened" (business as usual) a political impossibility. It reveals what had previously been hidden from view: that the sovereignty of the state had been effectively "bracketed off" (albeit with the state's tacit compliance) from an area of its legitimate and proper exercise. In the aftermath of disastrous failure, the unacceptable consequences of state subordination to the power of multinational corporations require confrontation with a new response.

In the case of offshore Gulf of Mexico, that response was a 6-month moratorium on deepwater drilling in U.S. coastal waters, announced by President Obama on May 27, the same day on which the head of the MMS tendered her resignation. Some 33 offshore drilling rigs ceased operations. The moratorium may be dismissed as a knee-jerk reaction to appease public opprobrium, but it ignited fierce opposition among the offshore industry with local political support as local jobs were threatened. Certainly it was duly lifted as the balance of geopolitical convenience (not least the rising price of gasoline for the American consumer) favored an early resumption in the eternal quest for energy security.

The far more acute question, however, is one of how regulatory regimes are reconstructed in the aftermath of regulatory failure. This goes to the heart of where the ultimate balance between the power of the state and that of capital lies. Whether and under what conditions it is possible "to bring the offshore back onshore" can best be addressed by looking at regulatory reconstruction and the attempt to "relegitimize" the regime in the aftermath of disasters on a similar scale. In this context, comparison with the aftermath of *Piper Alpha* is appropriate. In the UK offshore industry, following Lord Cullen's 1990 recommendations, a new fully resourced and inde- pendent regulator was established, with an authority to establish a new regime based on formal risk assessment and a safety case for each offshore installation. In the Gulf of Mexico, a new regime was also established. Any comments regarding this regime are necessarily preliminary given that its shape is still evolving. There are also serious objections that can be made based on the UK experience to claims that highly technical and confidential safety cases can in themselves provide a safety panacea in the reconstruction of the U.S. offshore regime (Steinzor, 2011; Woolfson et al., 1996). Nevertheless, there are some clues, at least as to potential pitfalls, which it might face based on the experience of the 20 years since *Piper Alpha*.

Just a few weeks before the lifting of the drilling moratorium in the last months of 2010, the Department of the Interior promulgated "new regulations on topics such as well casing and cementing, blowout preventers, safety certification, emer- gency response, and worker training" (National Commission, 2011: 152). In addition, a "new" U.S. regulator was established, the Bureau for Ocean Energy Management, Regulation and Enforcement (BOEMRE). To overcome the inherent conflict of purposes that had bedeviled the MMS, BOERME was itself reorganized in October 2011 into two new, independent agencies responsible for carrying out offshore energy management and enforcement functions: The Bureau of Safety and Environmental Enforcement (BSEE) and the Bureau of Ocean Energy Manage- ment (BOEM).

However, many oil and gas industry analysts complained that the change was only cosmetic, and the revamped agency was too close to its predecessor and lacked resources needed to hire new blood, especially of qualified inspectors and administrators. "The same people inside BOEMRE are the same people inside MMS. That ingrained culture is still the same. . . . You can't keep doing the same thing and expect different results," said Professor Bob Bea of the University of California's Deepwater Horizon Study Group (Christian Science Monitor, 2011). Even with the required funding and staff additions, it will take significant time for the new agencies to develop the capabilities to address the systemic risks associated with ultrahazardous hydrocarbon exploration and production projects (Deepwater Horizon Study Group, 2011: 94).

Equally as important as the immediate question of adequate resourcing and training of personnel for the new regulator, or its capacity to revamp offshore safety management practices, is a far more insidious process that quickly evidenced in the reconstruction of the safety regime in the UK offshore sector. This is characterized as "the gradual erosion scenario," whereby a legislative reform agenda mandating behavior change gradually deteriorates as it faces a host of "veto points." These emerge as concrete regulations are negotiated and amount to a renewed "strategy of containment" of regulatory interference and an attempt to reassert industry-led "compliance discretion." In the UK offshore sector, post–*Piper Alpha*, this particularly devolved to the industry's unwillingness to have the new safety case regime accepted by the new regulator (Woolfson et al., 2006: 361–365). In the U.S. sector, future research will determine whether and to what degree veto points have emerged, but even limited information available suggests that claims of an aggressive reform agenda in the new regulatory regime may be overstated (U.S. Department of the Interior, 2010). Prominent oil companies are already cautioning against "the rush to regulation." A new set of Workplace Safety Rules adopted in October 2010 now requires operators to have a comprehensive Safety and Environmental Management System (SEMS) that addresses the root cause of work-related accidents and offshore oil spills (National Commission, 2011: 258). The Workplace Safety Rule makes mandatory what was previously a voluntary but largely unimplemented program to identify, address, and manage safety hazards and environmental impacts. It was however devised by the American Petroleum Institute (API) as Recommended Practice 75. The continuing dependency of the new regulator on industry-set standards is herein exemplified.

Many observers of the industry rightly point out however that, in the UK sector of the North Sea at least, the new safety-case regime has been successful, as there has been no subsequent catastrophic event of fire and explosion similar to that of *Piper Alpha*. However, room for complacency would appear to be narrow. A review of this regime in 2007 by the Health and Safety Executive identified poor maintenance, cost cutting, managerial short-termism eroding operational safety and undermining commercial viability in which one third of tests for safety-critical items—fire pumps, deluge systems, temporary refuge heating ventilation and air conditioning (smoke protection)—scored red lights. Near-misses with potential consequences on the scale of a *Piper Alpha* have included the gas line rupture on BP's *Fortes Alpha* platform in 2003. This produced a major gas emission that fortunately failed to find a source of ignition, in part due to high winds that helped to disperse the gas cloud after about 20 minutes of extreme danger. The platform and its crew of 180 escaped unharmed. The incident resulted in one of the highest fines ever imposed on an offshore operator in the UK sector. Transocean, the owners of *Deepwater Horizon*, experienced a near miss on one of its North Sea rigs only 4 months before the Macondo well blow-out in a strikingly similar incident when gas entered the riser as the crew was displacing a well with seawater during a completion operation. Lessons from this incident were not transmitted to the Gulf of Mexico. Even in Norway, with its much-vaunted system of internal control, the Gullfaks incident in May 2010 provided a near-disaster scenario, which only luck prevented from being realized. In the words of the company itself, this event was due to "deficiencies in connection with risk management and compliance with internal requirements for drill operation planning and execution" (Statoil, 2010).

Finally, in late March 2012, a "well control incident" resulted in a gas release sufficiently serious as to require the evacuation of 209 personnel from the Total Elgin platform and a neighboring drilling rig in the Franklin field 150 miles off Aberdeen, Scotland (Upstream, 2012). With the platform enveloped in a flammable gas cloud, coastguards said shipping was being ordered to keep at least 2 miles away while a 3-mile air exclusion zone was imposed for aircraft. Shell also removed 120 nonessential staff from its two installations, about 4 miles from the Elgin because of concerns over drifting gas. At the time of writing, the platform had been powered down as the operators contemplated how to contain the release.

CONCLUSION

Three months after the *Deepwater Horizon* disaster, on July 30, BP established a $100 million charitable fund to assist rig workers experiencing economic hardships. However, this was dwarfed by the $20 billion fund that BP also created, at President Obama's urging, to compensate financial losses. In the first 8 weeks of the operation of the Gulf Coast Claims Facility, more than $2 billion was paid out to 127,000 claimants. Does BP's corporate contrition, including the change-out of its chief executive, signal a moment of epiphany on the long road to rebuilding a shattered corporate reputation and a renewed embrace of corporate responsibility? Equally, was BP the chief or the only corporate villain in the piece? What, if any, might be the appropriate sanctions that should apply in any event? Does the criminal justice system have a role to play here, not least in terms of a possible charge of corporate manslaughter? Should or could the "nuclear option" of corporate debarment and suspension (commonly referred to as "exclusion") be exercised (Steinzor, 2011)? A detailed discussion of these issues is beyond the scope of the present essay.

That BP exercised a long-standing corporate culture of cost cutting with tragic but foreseeable consequences seems beyond contest. In the case of *Deepwater Horizon*, the National Commission provided a detailed analysis of instances in which BP adopted drilling procedures or operational decisions that were designed to save time and therefore money—the use of "long string" well design, nonstandard procedures in the cementing process, the irregular use of centralizers and lockdown

202 / SAFETY OR PROFIT?

sleeves, deviations from normal operating procedure in the final processes of sealing the well (2011: 95–125). Yet BP operated the well in partnership with Anadarko and MOEX and employed as its key contractors Halliburton and Transocean. The National Commission (2011: 96) speaks of "a single overarching failure—a failure of management" and embraces in this all the key corporate partners. The Deepwater Horizon Study Group, under the leadership of the inestimable Professor Bob Bea, was clearer in apportioning the majority of blame on BP as the chief operator responsible, as indeed did the investigation of BOEMRE in its 2011 report. If BP was concerned with issues of safety, it was with the "high frequency, low consequence" events of the "slips, trips and falls" variety that present hazards to individual operatives. The more devastating "low frequency, high consequence" events as a result of a lack of a more systematic approach to processual risk seemed to have not received the same attention, said Bea (CBS News, 2010).

Without doubt, the attempt at mutual corporate blame-shifting as the executives of BP, Halliburton, and Transocean were collectively arraigned before congressional committees, presented a deeply unedifying spectacle that enraged the U.S. public from the president on down. Subsequent claims by BP of crucial evidence being destroyed by Halliburton only add to the sense of being unwilling witnesses to rather distasteful domestic feuding. Given the ultimate stakes however, in terms not only of which entity bears the current costs of compensation, but also future civil and criminal liabilities, such infighting was perhaps understandable. These included possible costs of $70 billion in future civil and criminal liabilities. In a plea bargain agreement with the Department of Justice, and in order to forestall a wider criminal indictment, in November 2012, BP agreed to a record $1.26 billion fine, the largest single criminal fine in U.S. history, as part of an unprecedented $4.5 billion settlement in which the company pleaded guilty to 14 criminal charges.

More important however than these outcomes is the eventual integrity of the new regime offshore. The "gradual erosion scenario" would suggest caution. One of the key conditions for the success of regulatory reform is the active support of all constituency groups (stakeholders). Yet the Achilles heel of the reconstruction of the new offshore safety regime is the lack of proposals that address the unresolved issue of employee empowerment. This is the final troubling similarity that exists between *Piper Alpha* and *Deepwater Horizon*. Ordinary workers in the industry who could have provided frontline safety intelligence and risk monitoring "from below" were afraid to voice their concerns (*New York Times*, 2010). In a hierarchically structured and even authoritarian system of management, employees were reluctant to speak out for fear of retaliation. As a result, employee voice was discounted, and this meant that the capacity of management to respond to early warning signals was already impaired. The oil companies and their contractor partners sought to create an offshore union-free environment, similar to many "offshore" export manufacturing and processing zones. In the UK offshore oil industry, a harshly anti-union atmosphere prevailed, and those who raised safety concerns were more often regarded as troublemakers, likely to be blacklisted by the employers (Woolfson et al., 2006: 143). Those who worked on the rigs off the coast of the Gulf of Mexico were also without union representation or collective voice (Corgey, 2010). A previous attempt to unionize the rigs at the start of the decade had been met with a vicious campaign orchestrated by a leading U.S. "union buster." It involved community agitation against union activists, best exemplified by hostile highway billboards carrying the warning, "There ain't no *you* in union" (ITF, 2002). Victimization, intimidation, and blacklisting remained rife in the so-called right-to-work Southern states.

The crucial dimension of employee empowerment in the safety management process was not to be adequately addressed directly by the President's National Commission or by other technical reports. Its absence presents a fundamental flaw in any attempted reconstruction of the safety regime. Corporate safety cultures may be revamped, but there is a finite limit to what can be achieved without the validation of employees' rights and their significant empowerment as key stakeholders with an independent voice in the safety management process. Only this, together with the necessary training to perform that role, can provide the crucial safety-auditing "from below" and the confidence and backup that enable employees to challenge managerial *diktat*.

As the search for oil offshore continues to be driven by the quest for energy security and the price per barrel reaches a new historic high, "offshore" will tend to remain just that, a "zone of regulatory exceptionalism." When the legitimacy of such arrangements is called into question, as in the aftermath of disaster, some redrawing of "territorial" jurisdictional boundaries between what is onshore and what is "offshore" takes place. "Sovereignty," at least in the sense of the power of the state to grant exemption, is rebalanced. New state regulators may emerge to reassert greater oversight, and readjustments may take place in the calculus between safety and profit, at least for a time. Ultimately however, the hegemony of the international oil industry and its ability to remold, subvert, and circumscribe the reach of national regulatory regimes threatens to nullify any attempted assertion of the prioritization of human life and environment over the bottom line. It has been argued here that the spatial/juridical "deterritorialization" of the regulatory regime in the offshore industry accounts for the "structuration of failure," exemplified by disturbing commonalities revealed in two major offshore oil disasters. Insofar as this industry pursues its objectives, sooner or later there will be new disasters to confront. Indeed, in one important sense, the hope must be that "offshore"—cost-driven, short-termist, and anti-union—does *not* return "onshore."

Afterword

Theo Nichols and David Walters

In retrospect, the decades immediately following the Second World War now have the appearance of an interregnum—a period of full employment in which labor strengthened and social protection in health, welfare, and work increased—which were then followed by a further dominance of market relations over social relations, with privatization and deregulation in many spheres. Writing in 1944, Karl Polanyi declared, "Undoubtedly our age will be credited with having seen the end of the self-regulatory market." Today, Polanyi's "double movement" (whereby free market excess brings forth social protection) has more the appearance, in Hyman's term, of a "double-double movement," whereby gains made in the immediate postwar period are themselves rolled back and the role of labor as a commodity is reasserted (Hyman, 2010: 309). Insofar as the organized working class is a key agent in the advancement of social protection and the fettering of market forces, its capacity is evidently now weakened by reduced trade union membership, by legal restraints on what trade unions may do (the prohibition of solidarity action in many countries, for example), and by a widespread decrease in bargaining power in the midst of high unemployment and underemployment.

It is in this general context that, in this Afterword, we examine the significance for health and safety at work of the position labor now occupies in the labor market and the nature of contemporary regulation under neo-liberalism. We also consider the limitations of so-called evidence-based policy as it is currently practiced and finally outline some priorities for future improvement.

CHANGE AND VULNERABILITY

A quarter of a century ago, one of us introduced the notion of "structures of vulnerability" in an attempt to explore the determinants of injury rates in UK manufacturing (Nichols, 1986). Three such components were held to be: unilateral control of health and safety by management, lack of trade union representation, and size of employment unit (small not being beautiful). Today, it would be remiss not to accord a more prominent place in such a list to the many deviations from what, at that time, would still have generally passed for the "standard model" of paid work, that of full-time employment with a single employer. Such a standard model is still well in evidence in many advanced capitalist societies. However, as Quinlan

206 / SAFETY OR PROFIT?

argues in Chapter 1, the rise of neoliberalism as the dominant policy framework in the 1970s saw the beginning of a shift in business practice toward outsourcing/ subcontracting, corporatization, privatization, downsizing/restructuring, and various forms of "lean" management. At the same time, changes began to emerge in the contractual basis on which workers were engaged—the term *precarious labor* being reinvented in the 1980s to encapsulate the various contractual relations that generally increased the vulnerability of those who work—temporary, agency, subcontract, self-employment, and other forms of labor, each of which variously impacts their health and safety. The important point about all such contractual relations is that they serve to exacerbate situations that give rise to the structures of vulnerability already described in more conventional forms of employment and at the same time act to further distance workers from the protective reach of organized labor. As Quinlan points out, the now quite extensive range of research that has explored this impact indicates that it acts to worsen the health, safety, and well-being of the workers concerned.

While women have always formed part of the labor force, one of the most significant features of the restructured labor market in developed economies in recent times has been the increased proportion of the labor force that are women. Yet, as Katherine Lippel and Karen Messing point out in Chapter 2, a gendered division of labor places them both at the lower levels of occupational hierarchies and in the performance by women of different types of work from men—even within work that bears the same job title. They further show that women are found more often than men in most precarious employment arrangements and how the hazards of the work undertaken by the majority of women are ignored or trivialized by all of the key actors in occupational health and safety.

A point repeatedly made by the authors of the previous chapters is that regulatory measures to protect workers from harm to their health and safety arising from pursuit of profit have been slow to adapt to these changes, whether in terms of gender sensitivity or more generally in relation to the increased precarity of work for both women and men. More than this however, a conclusion reached consistently across the range of jurisdictions represented in this book is that when adaptation has occurred, it has been to facilitate greater freedoms for employers to pursue profit than to benefit the protection of workers. In the next section, we offer some thoughts on the significance of this.

GOVERNANCE OF THE ECONOMY AND THE PLACE OF SAFETY AT WORK

From where we write, in the UK, a free-market ideology that depicts the "flexibility" necessary for the economic survival of the entire society is rampant. Indeed, such is the emphasis on the importance of business being free to act (including imposing the freedom of flexibility on those who work) that the regulation of health and safety has been largely presented by the current Conservative-Liberal coalition government as a "burden on business."

This view of regulation is not the exclusive domain of the Tory politicians who dominate the Coalition. It is important to be clear that, albeit to a lesser degree, it was a feature of the preceding New Labour Government, as epitomized in the creation of a special Department of Business Enterprise and Regulatory Reform. This view came to the fore under New Labour, following a brief period between its election in 1997 up until 2003, during which time an attempt had been made to restore the HSE to its pre-Thatcher era strength. The first of a series of budgetary restrictions on HSE resources was introduced in 2003. As Tombs and Whyte show in Chapter 5, as a result, HSE inspections fell, enforcement notices were issued less often, and prosecutions fell. The current Coalition government differs only in the scale and passion with which it attempts to outrun the worldwide tide of neo-liberalism. It differs in its degree of commitment to the idea that the state must be shrunk; in its conviction that regulation is inimical to enterprise, growth, and profit; and not least in the extent to which it seeks to tar all regulation of health and safety with the seemingly asinine examples of "overregulation" that it delights in putting before the public.

The UK was, of course, the birthplace of laissez faire, and neither neo-liberalism nor, for that matter, the earlier Robens philosophy, are a far step from this. On the face of it, however, the Coalition's neo-liberal approach takes on the appearance of being diametrically opposed to the ideology that underpinned the Robens Report. This was that safety and profit went together; so that, as Robens put it, "there is greater natural identity of interest between 'the two sides' in relation to health and safety problems than in most other matters" (Robens, 1972: para. 66). Yet, as Tombs and Whyte have noted, whereas the Robens and business case on health and safety was based upon claims for profits *and* safety (that the two went together), the opposite is now being proposed—that we can have safety *or* profits, so that financial recovery actually demands less regulation. Since the advent of the 2010 Coalition government, there has also been a change of emphasis that relates to this, which concerns the centrality of workers. Robens referred to the two sides of industry as the "two sides," so as to cover his embarrassment at the very thought that industry might be a site of conflict, but at least he referred to them. By contrast, the focus of the neo-liberal rhetoric is now one-sided; it is on the entrepreneur (employer/manager), especially upon the burden that red tape puts on the entrepreneur (the so called Red Tape Challenge); on the paperwork and on unnecessary procedures that hold the entrepreneur back and that stop the cornucopia overflowing to the common benefit. Workers? What workers? They scarcely figure, and certainly not as part of the solution to improved health and safety. In fact, of course, the issue, for the government, is not health and safety; it is profit and the alleged barriers to its augmentation.

Currently in the UK, then, despite some affinity with the idea of "self-regulation," which was central to the Robens Report (1972), the neo-liberal rhetoric on health and safety at work is not driven by an overwhelming concern with the health and safety of workers: it is driven by an overwhelming concern with the economic welfare of the supposed real risk takers, the entrepreneurs, who must be freed from the Nanny State and its petty restrictions. Consistent with the tone of this is the rhetorical flourish of the Chancellor of the Exchequer in the 2012 budget that the government would scrap or improve 84% of health and safety regulation. However, this apparent root and branch war on health and safety regulation, as suggested by the 84% figure, is perhaps not all that it seems. It covers alleged "improvement" and not only scrapping. More significantly, it will doubtless entail acting on legislation that is outmoded by virtue of having been on the statute book for many years (and the removal of which, it might be thought, is also unlikely to do anything to stimulate enterprise). For instance, the regulations cited by the 2011 Löfstedt Report (2011: 104–106), which recommended regulatory reform and which had been commissioned by the government to "consider the opportunities for reducing the burden of health and safety legislation on UK businesses whilst maintaining the progress made in improving health and safety outcomes" included cases that were first introduced 1875, with a considerable number of cases relating to mining and quarrying, first introduced over half a century ago. Löfstedt himself made clear subsequent to the publication of the Report (SHP, 2011), these changes are to be distinguished from the government's "war on health and safety."

The thrust of government policy is evident from the government's response to the Löfstedt Report (2011). Employers are to get more legal protection: all regulatory provisions that impose strict liability will be removed in an attempt "to address what could be a significant driver of over-compliance with health and safety law" (DWP, 2011: 91). The self-employed (a burgeoning number of whom have been driven to adopt this status out of economic necessity) will be exempted from "health and safety burdens" in cases where they are in low risk occupations whose activities represent no risk to other people (DWP, 2011: 9). Inspection will be refocused on higher risk areas; it will be lightened for lower risk businesses who manage their responsibilities effectively (DWP, 2011: 6). All proposed EU directives and regulations (and amendments to them) that have a perceived cost to society of more than €100 million should go through an automatic regulatory impact assessment (DWP, 2011: 14).

It is clear from the various chapters in this book that state regulation—in the form of setting financial penalties for work health and safety offenses, prosecuting employers, and imposing fines—has varied between countries. In Britain, Tombs and Whyte document a decline in inspection, investigation, and enforcement action such that they conclude, "Robens' original plan to replace external regulation with self-regulation has come to pass." In the same vein, but in relation to occupational health, Watterson notes in Chapter 7 that successive UK governments have run down the provision of state occupational health staff and services while at the same time adopting the so-called better regulation agenda. He argues that in practice, this has meant softer regulation or no effective external regulation at all for many of the situations known to cause such diseases as occupationally related cancer. He points out also that this weakness is linked with minimal sanctions against employers responsible for exposing workers (O'Neill et al., 2007).

In Chapter 6, Johnstone reports on major initiatives in Australia to significantly increase the level of maximum fines, to introduce a number of nonfinancial sanctions such as modified community service orders (and in one state, Australian Capital Territory, the implementation of corporate manslaughter reforms) but he sees little sign of the view being overcome that health and safety offenses are not "really criminal."

In Ontario, Tucker reports in Chapter 4, there was until recently an intensification of enforcement activities, though he again notes little use of criminal sanctions. In Chapter 8, Lewchuk, also writing on Ontario, notes an acceptance by policy-makers that asking precarious workers to self-regulate is not likely to work, but that this certainly does not signal a return to a regulatory regime based on external enforcement of health and safety standards and criminal prosecutions of employers who fail to protect the health of workers, as preferred by Ontario labor. The changes in Canada to the structure and organization of work, resulting from the opening of markets to greater international competition, can be seen to be similar to those observed in Europe. As elsewhere, they can also be seen to have broadly the same consequences for the deregulation of health and safety and regulatory failure to account of the greater vulnerability of a workforce that has been made more precarious by these changes.

Generally, the direction of travel of the governments of the other countries referred to in the previous chapters is not dissimilar from that of the UK. The feeble American regulation of offshore drilling in the Gulf of Mexico was a yet more pronounced version of the widely favored "soft"/"light touch"/deregulatory approach, as Woolfson shows in his account of the *Deepwater Horizon* disaster in Chapter 9. There, the regulatory regime was such that external facilitation of "energy security" rendered safe working secondary to collecting revenue for the U.S. Treasury, a situation well described as "regulatory capture," and a blatant example of the triumph of profit over safety. In this version of safety regulation, the federal government, in the guise of the Mineral Management Service, ceased to function as even the corrective backstop envisaged in the most minimalist versions of the Robens model (1972) or the related New Governance theory. What Carson (1982), in his earlier seminal analysis of safety in the North Sea, had called "the political economy of speed" took over. As was the case with the 1988 *Piper Alpha* disaster in the North Sea, as Woolfson states, "Everything pointed to haste, corner cutting, cost saving and to sloppy management."

The EU is often depicted by the current UK government as the source of regulations that fetter enterprise and burden business with red tape. But the impression this can give, of a pro-regulatory EU, is misleading. The reality is that to the extent that the EU did once favor not only the free movement of people and capital but the development of a "Social Europe," in which there would be a social market economy with improved working conditions and an improved standard of living for workers, the latter objective has faded from prominence. As Vogel and Walters (2009) have previously outlined, in the several decades of supranational regulation of occupational health and safety in the European Union, the character of regulatory policy has been shaped by the same wider political and economic determinants as those influencing its character and progress at the national level. For a brief period, this meant that the character of community regulatory policy was undoubtedly influenced by the vision of a social Europe associated with Jacque Delors. The same period coincided with and influenced the spread of the process-based reflexive regulation that currently styles the approach to OHS regulation in Europe and in most developed economies internationally (see Walters, 2002). However, the European Community approach has changed fundamentally since that time.

In a series of well-documented policy moves and treaties from the Maastricht Treaty of the early 1990s onward, the regulatory policies of the EU and their administration through the Commission have increasingly reflected the wider influence of the free-market rhetoric and the interests behind it that drive economic globalization. A consequence of this is seen in the growing abrogation of responsibility for effectively regulating the protection of the health, safety, and well-being of millions of workers in the European Union. Essentially, the same promotion of profit we have described as determining current UK governance of OHS has effectively overtaken that of the protection of worker safety in EU policymaking too.

Under the EU version of "new governance," and in line with the supposed economic wisdom espoused by international bodies like the Organization for Economic Co-operation and Development (OECD), the interests of capital are increasingly pervasive, undermining the concepts of social protection under which the agenda for OHS policy was previously conceived. The approach is consistent with the notion that public regulation of any kind is a burden on business growth, and therefore its legitimacy must be measured by economic impact assessments, cost benefit analysis, and the like; the methods for which are based around a set of assumptions biased toward the short-term interests of capital (see, for example, van den Abeele, 2009; Verheugen, 2007; Vogel, 2010).

The approach is, of course, not restricted to regulating OHS, indeed, it is arguably not primarily aimed in this direction, but part of a wider deregulatory agenda pursued by the Commission (EC, 2009). However, preventive measures for OHS are especially vulnerable. For example, one of the positions adopted by the Commission, which, as we have seen, is also familiar at national levels, is that culling legal requirements on employers to provide information can reduce administrative burdens on business. As Vogel (2009) has argued, such requirements to provide information are fundamental to the operation of the model of health and safety management that was implemented throughout the EU by the adoption of the 1989 Framework Directive 89/391. Attempts to dilute requirements on employers in relation to workplace risk assessment, on the provision of information on injuries, incidents, work-related ill-health, and working time—as well as the ongoing sophistry (well illustrated in the current EC Better Regulation agenda) that small and medium-sized firms must be protected from "excessive regulation"—have been a continued feature of Commission policy positions over the past decade (EC, 2009). As Vogel (2009) has eloquently shown, they act to undermine the effectiveness of the participatory approaches to OHS management that the Directive itself requires.

As a result of the shift in policy orientation, there is now a well-established brake placed on the introduction of new Directives on health and safety, and there have been successive attempts to remove or water down the requirements of existing Directives through the advice and actions of various groups at the community level, such as those of the so-called "High Level Group of Independent Stakeholders on Administrative burdens" (EC, 2007). Through the use of private consultants and spurious analysis borrowed from member states such as The Netherlands and the UK, as well as the OECD, the Commission has adopted models of impact assessment and cost-benefit analysis to be applied both to the possible introduction of new regulations as well as in the operation of existing ones (EC, 2005, 2006; van den Abeele, 2010: 25–26).

These changes are further reflected in the shift that has occurred in the tone and content of successive Community Health and Safety at Work Strategies since the 1980s, when the last of Delors' vision of a Social Europe led to the adoption of the Framework Directive 89/391 and its daughter directives. With each successive strategy, greater emphasis has been placed on the business case for health and safety and the role of softer options for its implementation. There has been increasing emphasis, not on the need to protect European workers but on the notion that a healthy and safe workforce is likely to be more profitable. At the time of writing, the position has moved so far that a major element of the debate around the creation of a new Community Health and Safety at Work Strategy for 2012 onward which is already much-delayed, has concerned whether it is necessary to have a strategy at all (see, for example, SLIC, 2012).

At the same time, alternatives to regulation have been sought and initiatives involving "soft law" have replaced regulatory actions on a variety of subjects. The argument has been advanced that, in some cases, they serve to achieve more workable solutions to problems of industrial relations and labor regulation that have proved intractable to regulation (Bercusson, 2008). However, as Bercusson also pointed out, there are considerable uncertainties as to whose interests are actually best served by many of these alternatives and the administrative means chosen to facilitate them. Here again, the one thing that stands out in this lexicon of "new governance" in Europe, is the primacy of the aim of states, the community, and the Commission to achieve an enhanced business environment for employers, which takes precedent over concerns over the health and safety of workers.

There is then, a substantial degree of harmony between thinking at the level of the European Union and that of governance of member states like the United Kingdom, long regarded to be somewhere toward the extreme end of the free-market spectrum in ideological terms. However, the same effects are found at national levels in other member states too. Perhaps most tellingly, they are seen even in countries once considered to be models of enlightenment in relation to social policy. To appreciate the extent of the pervasive influence of the neo-liberal ideology on OHS policy, we can do no better than to consider developments in Sweden, once regarded as a paragon of social democratic approaches to managing health and safety at work.

As Frick details in Chapter 3, the last decade has witnessed an unprecedented attack on the underpinnings of the consensus style management of the work environment that had been the leitmotif of the "Swedish model" of work environment

212 / SAFETY OR PROFIT?

regulation in the postwar era. This model was grounded in a strong economy and labor market in which there was high employment and where labor was therefore able to occupy a powerful position to negotiate work environment norms with employers and the state. A weakened economy and rising unemployment has led, in turn, to relatively weaker trade unions along with increasing cost cutting by the state and, in line with most other advanced market economies, a search for neo-liberal political and economic solutions to improving the business environment for employers, often at the expense of measures to support the health and safety of workers. Thus, the resourcing of the Work Environment Authority was drastically reduced, and the frequency of inspections has fallen as a result. Research and support for good practices on the work environment has been substantially diminished since the government's closure of the National Institute for Working Life in 2007, a beacon of European research into health and safety at work.

At the same time, also mirroring international trends, the structure and organization of work in Sweden has shifted from one in which there was a dominance of large organizations in manufacturing, to a service-based and increasingly privatized and outsourced model, while the state has removed or reduced regulatory requirements to help facilitate this change. In parallel, it has also increased the cost of belonging to a trade union and reduced eligibility for social welfare benefits for those who are not in work. These are familiar developments when viewed from a UK perspective, and as in the UK, all are directed at easing the economic path of employers while ignoring the needs and protection of their workers.

AN EVIDENCE-BASED APPROACH TO POLICY?

In recent years, calls for evidence-based policy have become commonplace. Like much of the "commonsense" approach to regulating risks and protecting the health and safety of workers, at first sight, this seems like a good idea. But we need to be clear on what is meant by this term and consider its usage in practice, because in practice, the idea that policy is driven by the search to enhance health is perverted.

"Evidence" in occupational health and safety has long been contested territory. As Harremoës, Gee, and MacGarvin (2001) show in their painstaking compilations of historical and current case studies concerning the late application of precautionary principles in state prevention policies, time and time again, this has resulted in needless human suffering and loss of life, often to the benefit of the business and the profits of companies creating the risks to health. In their detailed accounts of the relationship between scientific evidence and policy action, whether addressing risk prevention in relation to occupational or environmental exposures or in protecting consumers or patients, they present a host of examples of situations in which risk-prevention policies have lagged far behind the warnings of emergent scientific evidence concerning risk. In terms of occupational risks, this was true, for example, in the production and use of asbestos, lead, vinyl chloride, benzene, polychlorinated biphenyls, isocyanates, and many other now commonly accepted hazardous materials. It has been shown to be equally true in cases involving major disasters in which financial computations effectively delayed or prevented managerial action on known risks, as we saw in Woolfson's account of the BP *Deepwater Horizon* disaster. All these examples show that in complex and uncertain risk scenarios, ethical judgments concerning the strength of evidence are often ultimately necessary, and in these judgments, short-term economic interest needs to be distinguished from the longer-term health of people and their environment. In other words, a distinction is required between the desirability of health or profit. An alarming feature of current political interpretation of evidence-based risk-prevention policies is that the balance of this distinction has shifted in ways that favor the latter, as amply illustrated, for example, by the regulatory impact analysis now increasingly demanded in policies on regulation in the European Union and among its member states, which effectively require a cost-benefit analysis of spurious quality that is biased toward economic rather than health considerations (Vogel, 2009).

Testimony to the fact that the assessment of evidence is not as straightforward as sometimes implied are the British figures on work-related fatalities, which constitute a very considerable underestimate of the extent of work-related harm, as Tombs and Whyte show in Chapter 5. Indeed, whereas there may be some case for the retention of official time series to monitor changes in fatal injuries, on the grounds (themselves in need of empirical verification) that they measure like with like, there is no excuse to allow them to stand as unqualified valid indicators of the extent of overall work-related harm, which in effect they do, for example, through press releases that announce that over a given year the number of fatalities has fallen to "yet another record low."

In the case of work-related disease rather than event-specific fatality or injury, matters are still less straightforward, as Watterson shows in Chapter 7 with reference to occupational cancer. Estimates of the extent of cancers resulting from occupational exposures have long been disputed, however there is no dispute that even by conservative estimates, occupationally related cancer mortality is a significant element of occupational mortality overall. Moreover, as Watterson comments, when risks are established, as is the case in relation to many carcinogenic exposures, be they chemical, physical, biological, or caused by the organization of work, precautionary policies would suggest that preventive action should be relatively obvious. This is not the case, largely because of complex and competing vested interests that interact in the UK in ways resulting in the continued absence of an effective occupational cancer policy on the part of the state and its regulators. Dominant in these interests are those of capital, but as Watterson shows, its dominance often operates in ways that capture others and help to ensure the current peculiar position of policymakers and regulators who produce reports on the economic costs of occupational diseases in the UK, but who then effectively ignore their own analyses in subsequent action (or more to the point, inaction) on policy.

In Chapter 8, a further illustration of the limits to the so-called evidence-based approach is found in Lewchuk's account of a 2010 regulatory review in Ontario, set up, among other things, to examine the continued relevance of regulatory provision in the light of the spread of precarious employment. He shows how it was the wider political context that influenced the findings of the Review Panel as much as it was the detail and quality of the evidence submitted to it. In the case of the latter, he further shows that this detail and quality was far greater and more wide-ranging in the case of the submissions from labor than was so in those from employers groups. Yet, largely in deference to pressures from the wider political context, in its report, the Review Panel focused on issues of cost and safety behavior that had dominated the employers' limited submissions rather than on the more fundamental reforms argued in detail by labor. The result, in Lewchuk's view, was a set of recommendations heavily influenced by the same "safety or profit" constraints to which we have already referred and which would go only a small way to deal with the issues confronting the increasingly precarious workforce whose needs the Panel had set out to address. What is needed, Lewchuk argues, is a culture change that shifts the balance between profits and safety and puts greater value on allowing workers to work without shortening their lives.

It is precisely because the evidence on health is so contested that worker representation becomes such an important issue. It is therefore fitting we end this chapter with some reflections on its current role.

WHAT IS TO BE DONE?

Some 40 years ago, in the eponymously titled booklet to which all the authors of the previous chapters have referred, one of us wrote, "If human costs are to be minimised it is production itself that must be controlled" and "any fundamental and solidly based improvement . . . must depend on a shift of power to control production to those who are now getting hurt" (Nichols and Armstrong, 1973: 33). Nothing in the intervening years has caused either of us to doubt this conclusion, and much in the present volume explains why this is so. However, what has occurred during this time, as we outlined at the beginning of this chapter has been a substantial shift in the balance of power between labor and capital, which has served to make is considerably more difficult to achieve this desired control. Despite the present challenge, however, at least in the early part of the intervening years since the publication of the booklet, substantial gains were made by labor in its efforts to instigate institutional arrangements with which it might potentially achieve such a shift. National regulatory measures, international requirements, as well as ILO Conventions have all acknowledged the right of labor to be represented and consulted on health and safety matters at the workplace, at sector, national, and international levels. Institutional arrangements exist at all these levels to effect this. Moreover, the evidence, such as it is, demonstrates that they work. They play a role in improving health and safety outcomes, and as we have shown in previous work, there is little doubt that health and safety performance outcomes are better when health and safety is managed in the presence of arrangements for worker representation and consultation, than when employers attempt to manage it in their absence.

While it is clear that many such arrangements have been usurped or accommodated by capital, and their impact has been substantially eroded by wider changes in the structure and organization of work and in the position of labor, it remains the case that, by and large, these institutional arrangements have been left in place and not dismantled by the efforts of neo-liberal governance.

In the UK, for example, despite the neo-liberal affinities of the 2010 Coalition government and previous Labour administrations, the sphere of health and safety regulation remains the only one in which tripartite governance by employer, employee, and state representatives persists. Similarly, the right of trade union representatives to call for the establishment of health and safety committees remains intact. Elsewhere, too, important elements of institutional structure have survived the neo-liberal and pro-employer tide. In Sweden, despite the fact that one of the first actions of the incoming 2006 government was to abolish the National Institute for Working Life and the other inroads it has made, it remains the case that the deterioration in cooperation over the work environment and in the number and influence of safety representatives is less than might be expected, as Frick reports in Chapter 3. In Australia, despite the attempt to decriminalize the law, it has been the case in the recent past that there has been a major debate about improving the enforcement of work health and safety statutes. In Canada, both the internal responsibility system and joint health and safety committees received significant support from the 2010 Dean Panel review.

These positive developments do little to shift the main reality. For the effectiveness of worker representation and consultation does not depend solely on the existence of regulatory measures or even on the appointment of representatives and joint committees. There are a set of preconditions that determine the effectiveness of health and safety arrangements, which include the role of regulatory inspection; strong trade union organization, both within workplaces and outside them; commitment from employers to both health and safety; and a participative approach to achieving improvements, as well as well-trained worker representatives and strong links between them and their constituents (see Walters and Nichols, 2007: 117). The tendencies pointed to in this book erode all these preconditions, and there is little hope that such institutional gains will deliver anything like their potential in the political climates prevalent in developed economies at the present time. However, they demonstrate what is possible when circumstances allow organized labor to assert its demands, and they remain structures to be defended.

Organization and resistance to exploitation is not at an end. As Tucker argues in Chapter 4, while "new governance" theory (an upmarket version of an old conventional wisdom) may offer little, others have pointed to possible emergent groupings in civil society that could act by themselves or in concert with public enforcement bodies to help give voice to more vulnerable workers. Writers such as Janet Fine (2006) and Milkman, Bloom, and Harro (2010: 1–20) have commented in relation to the situation in the United States, where organized labor is especially weak and large groups of workers lie beyond its influence, that alternative forms of organization, among immigrant workers for example, offer new ways of addressing worker rights on health and safety.

These are, of course, small isolated cases in a scenario in which the opportunities for workers to assert their collective voice on health and safety are increasingly constrained by the power of capital, but they are illustrative of the ongoing struggle of workers in the face of the unilateral interests of capital, to organize collectively to give voice to their demand for protection of their health, safety, and well-being by whatever means available. The importance of their voice in this respect cannot be questioned. As Woolfson notes in Chapter 9, in cases of catastrophic failures such as *Deepwater Horizon* and the earlier *Piper Alpha* disaster, operatives had voiced safety concerns over a period of time that management failed to respond to adequately or identify as warning signals that their systems were deficient. Similar conclusions have been reached time and time again in the aftermath of such events. In each case, the absence of effective channels for workers' collective voice to be heard greatly increased the risks of catastrophe. Today, in the midst of high unemployment and with labor weakened versus capital, it is yet more evident than 40 years ago that whether it be men or women, in factory, field, office, or other workplace, and whatever their contract of employment "any fundamental and solidly based improvement . . . must depend on a shift of power to control production."

References

AFS. (2001). *Systematiskt Arbetsmiljöarbete. Arbetsmiljöverkets Författningssamling 2001:1.* Solna, Sweden: Arbetsmiljöverket.

Ahlgren, C., Olsson, E.-B. M., and Brulin, C. (2012). Gender analysis of musculoskeletal disorders and emotional exhaustion: Interactive effects from physical and psychosocial work exposures and engagement in domestic work. *Ergonomics, 55*(2): 212–228.

Aiken, L. et al. (2002). Hospital nurse staffing and patient mortality, nurse burnout and job dissatisfaction. *Journal of American Medical Association, 288*: 1987–1993.

Ala-Mursula, L., Vahtera, J., Linna, A., Pentti, J., and Kivimaki, M. (2005). Employee worktime control moderates the effects of job strain and effort–reward imbalances on sickness absence: The 10-town study. *Journal of Epidemiology and Community Health, 59*: 851–857.

Aldrich, M. (1997). *Safety first.* Baltimore, MD: Johns Hopkins University Press.

Alliance for Cancer Prevention. (2012). *Campaigning and prevention.* Retrieved February 22, 2012 from http://www.allianceforcancerprevention.org.uk/

Arbetarskydd. (2000). Fyra av fem arbetsplatser saknar skyddsombud. *Arbetarskydd, 5.*

Arbetarskydd. (2003). Skyddsombuden ökar igen. *Arbetarskydd, 4*: 5–7.

Arbetarskydd. (2004). Få skyddsombud i privata tjänster. *Arbetarskydd, 3*: 4–5.

Armstrong, P., and Armstrong, H. (1994). *The double ghetto.* Toronto, Canada: McClelland and Stewart.

Armstrong, P., Banerjee, A., Szebehely, M., Armstrong, H., Daly, T., and Lafrance, S. (2009). *They deserve better: The long-term care experience in Canada and Scandinavia.* Ottawa: Canadian Centre for Policy Alternatives.

Armstrong, A., Dauncey, G., and Wordworth, A. (2007). *Cancer: 101 solutions to a preventable epidemic.* British Columbia, Canada: New Society Publishing.

Aronsson, G., Dallner, M., Lindh, T., and Goransson, S. (2005). Flexible pay but fixed expense: Personal financial strain among on-call employees. *International Journal of Health Services, 35*(3): 499–528.

Aronsson, G., Gustafsson, K., and Dallner, M. (2000). Sick but yet at work: An empirical study of sickness presenteeism. *Journal of Epidemiology and Community Health, 54*: 502–509.

Arthurs, H. (2009). Corporate self-regulation: Political economy, state regulation and reflexive labour law. In C. Estlund and B. Bercusson (Eds.), *Regulating labour in the wake of globalization: New challenges, new institutions.* Oxford, UK: Hart.

Asselin, S. (2003). Professions: Convergence entre les sexes? *Institut de la statistique: Données sociodémographiques en bref, 7*(3): 6–8.

Association of Scientific Technical and Managerial Staffs (ASTMS). (1980). *The prevention of occupational cancer.* London, UK: ASTMS.

ATK. (2000). *Bättre möjligheter till en bättre arbetsmiljö—Om skyddsombudens informations—och kunskapsförsörjning. Behov—möjligheter – tillgång – utbud.* Stockholm, Sweden: Arbetstagarkonsult AB.

218 / SAFETY OR PROFIT?

Australian Law Reform Commission (ALRC). (2002). *Report 95: Principled regulation: Federal civil and administrative penalties in Australia.* Sydney: Commonwealth of Australia.

AV. (2010). *Arbetsmiljö 2009.* Stockholm, Sweden: Arbetsmiljöverket.

AV. (2011a). *Årsredovisning 2010.* Stockholm, Sweden: Arbetsmiljöverket.

AV. (2011b). *Rapport partsmedel 2010.* Stockholm, Sweden: Arbetsmiljöverket.

Ayres, I., and Braithwaite, J. (1992). *Responsive regulation: Transcending the deregulation debate.* New York, NY: Oxford University Press.

Baker, K. (2009). 7,000 people die each year from avoidable work-related cancers. *Personnel Today.* Retrieved March 10 from http://www.personneltoday.com/articles/2009/03/10/49753/7000-people-die-each-year-from-avoidable-work-related.html

Bardach, E., and Kagan, R. (1982). *Going by the book: The problem of regulatory unreasonableness.* Philadelphia, PA: Transaction.

Barling, J., and Mendelson, M. (1999). Parents' job insecurity affects children's grade performances through the indirect effects of beliefs in an unjust world and negative mood. *Journal of Occupational Health Psychology, 4*(4): 347–355.

Barthe, B., Messing, K., and Abbas, L. (2011). Strategies used by women workers to reconcile family responsibilities with atypical work schedules in the service sector. *Work, 40*(Supplement): S47–S58.

Bartrip, P. W. J., and Fenn, P. T. (1983). The evolution of regulatory style in the nineteenth century British factory Inspectorate. *British Journal of Law and Society, 10*(2), 201–222.

BBC. (2010). *BP boss Tony Hayward's gaffes.* June 20. http://www.bbc.co.uk/news/10360084

BBC World News America. (2011). *BP oil spill commission's William Reilly on what went wrong.* Retrieved January 7 from http://news.bbc.co.uk/2/hi/programmes/world_news_america/9349780.stm

BBIS. (2011). *Svart jobb—vilket Mörker! Byggbranschen i samverkan.* Available at http://www.renbyggbransch.nu/

BC Court. (2011). Union's private criminal negligence case can go forward. *OHS Insider.* Retrieved August 8, 2011 from http://ohsinsider.com/search-by-index/c45/bc-court-union%E2%80%99s-private-criminal-negligence-prosecution-can-go-forward

Benach, J., Muntaner, C., and Santana, V. (Eds.), and Employment Conditions Knowledge Network (EMCONET). (2007). *Employment conditions and health inequalities: Final report to the WHO Commission on Social Determinants of Health (CSDH).* Geneva, Switzerland: World Health Organization.

Benach, J., Muntaner, C., with Solar, O., Santana, V., and Quinlan M. (2010). *Empleo, Trabajo Y Desigualdades en Salud: Una Vision Global,* Barcelona, Spain: University of Pompa Fabra.

Bennie, I., Bachl, B., and Dewey, R. (2010). Success, one small step at a time. *CAW Health, Safety Environment Newsletter,* September–December.

Bercusson, B. (2008). A changing institutional architecture of the European social model. In B. Bercusson and C. Estlund (Eds.), *Regulating labour in the wake of globalisation: New challenges, new institutions.* Oxford and Portland, OR: Hart Publishing.

Bernstein, S. (2006). Mitigating precarious employment in Quebec: The role of minimum employment standards legislation. In L. Vosko (Ed.), *Precarious employment: Understanding labour market insecurity in Canada* (pp. 221–240). Montreal, Canada: McGill-Queens University Press.

REFERENCES / 219

Bernstein, S., Lippel, K., Tucker, E., and Vosko, L. (2006). Precarious employment and the law's flaws: Identifying regulatory failure and securing effective protection of workers. In L. Vosko (Ed.), *Precarious employment: Understanding labour market insecurity in Canada* (pp. 203–220). Montreal, Canada: McGill-Queens University Press.

Bittle, S. (2013). *Still dying for a living.* Vancouver, Canada: University of British Columbia Press.

Bittle, S., and Snider, L. (2006). From manslaughter to preventable accident: Shaping corporate criminal liability. *Law & Policy, 28*: 470–498.

Black, J. (2001). *Managing discretion.* Paper presented at the Australian Law Reform Commission Conference on Penalties: Policies and Practice in Government Regulation, Sydney.

Black, J. (2010). Risk-based regulation: Choices, practices and lessons being learnt. In G. Bounds and N. Malyshev (Eds.), *Risk and regulation policy. Improving the governance of risk.* Paris, France: OECD.

Blackett, A. (2011). Regulating decent work for domestic workers. *Canadian Journal of Women and the Law, 23*(1): 1–45.

Bluff, E., and Johnstone, R. (2003). Infringement notices: Stimulus for prevention or trivialising offences? *Journal of Occupational Health and Safety—Australia and New Zealand, 19*: 337–346.

Bohle, P., Pitts, C., and Quinlan, M. (2010). Time to call it quits: Older workers, contingent employment, safety and health. *International Journal of Health Services, 40*(1): 23–41.

Bohle, P., Willaby, H., Quinlan, M., and McNamara, M. (2011). Flexible work in call centres: Working hours, work-life conflict and health. *Applied Ergonomics, 42*(2): 219–224.

Boivin, D., Tremblay, G., and James, F. (2007). Working on atypical schedules. *Sleep Medicine, 8*: 578–589.

Bolinder, E., Magnusson, E., Nilsson, C., and Rehn, M. (1982). *Vad händer med arbetsmiljön?* Stockholm, Sweden: LO and Tidens förlag.

Brabant, C., Mergler, D., and Messing, K. (1990). Va te faire soigner ton usine est malade: La place de l'hystérie de masse dans la problématique de la santé des travailleuses. *Santé mentale au Québec, 15*(1): 181–204.

Braithwaite, J. (2002). *Restorative justice and responsive regulation.* Oxford, UK: Oxford University Press.

Braithwaite, J. (2008). *Regulatory capitalism.* Cheltenham, UK: Edward Elgar Publishing.

Braithwaite, J. (2011). The essence of responsive regulation. *UBC Law Review, 44*(3): 475.

Brophy, J. T., Keith, M. M., Watterson, A., Park, R., Gilbertson, M., Maticka-Tyndale, E., et al. (2012). Breast cancer risk in relation to occupations with exposure to carcinogens and endocrine disruptors: A Canadian case-control study. *Environmental Health 11*: 87. doi: 10.1186/1476-069X-11-87

Brown, G. (2005). A plan to lighten the regulatory burden on business. *Financial Times,* May 23.

Bruhn, A., and Frick, K. (2011). Why it was so difficult to develop new methods to inspect work organization and psychosocial risks in Sweden? *Safety Science, 49*: 575–581.

Bureau of Labor Statistics. (2009). *Women in the labor force: A databook.* U.S. Department of Labor.

Bureau of Ocean Energy Management, Regulation and Enforcement (BOEMRE). (2011). *Report regarding the causes of the April 20, 2010 Macondo well blowout. Final report of joint investigation team.* Retrieved September 14 from http://www.boemre.gov/pdfs/maps/DWHFINAL.pdf

220 / SAFETY OR PROFIT?

Buxton, O., Quintiliani, L., Yang, M., Ebbeling, C., Stoddard, A., Pereira, L., and Sorensen, G. (2009). Association of sleep adequacy with more healthful food choices and positive workplace experiences among motor freight workers. *American Journal of Public Health, 99*: S3, S636–S643.

Cabinet Office. (2006). *New bill to enable delivery of swift and efficient regulatory reform to cut red tape—Jim Murphy*, Cabinet Office News Release, January 12, 2006. London, UK: Cabinet Office Press Office. Available at http://www.egovmonitor.com/node/4164

Calvet, B., Riel, J., Couture, V., and Messing, K. (2012). Work organization and gender among hospital cleaners in Quebec after the merger of "light" and "heavy" work classifications. *Ergonomics, 55*(2): 160–172.

Cameron, D. (2009). http://news.bbc.co.uk/1/hi/8388025.stm. Retrieved October 14, 2011 from http://www.telegraph.co.uk/news/politics/conservative/8808521/Conservative-Party-conference-2011-David-Camerons-speech-in-full.htm

Cameron, D. (2012). PM vows to "kill off the health and safety culture for good." *Safety and Health Practitioner*, February.

Campolieti, M., Hyatt, D., and Thomason, T. (2006). Experience rating, work injuries and benefit costs: Some new evidence. *Relations Industrielles, 61*: 118–145.

Canadian Employment Law Today. (2011). C-45 Charges against Ontario crane company dropped. *Canadian Employment Law Today* (January 4, 2011). Retrieved August 8, 2011 from http://www.employmentlawtoday.com/ArticleView.aspx?l=1& articlcid=2450

Cancer Research UK (CRUK). (2009). Denmark compensates women who develop cancer after working night shifts. Cancer Research UK news release, March 18, 2009.

Carlin, J., Knight, J., Pickvance, S., and Watterson, A. (2008). A worker-driven and community-based investigation of the health of one group of workers exposed to vinyl chloride monomer (VCM). *Journal of Risk and Governance, 1*(2): 1–18.

Carpentier-Roy, M.-C., and Vézina, M. (Eds.). (2000). *Le travail et ses malentendus: Psychodynamique du travail et gestion: Enquêtes en psychodynamique du travail au Québec.* Saint-Nicolas, Canada: Presses de l'Université Laval (Les)/Octares.

Carson, W.-G. (1979). The conventionalisation of early factory crime. *International Journal of the Sociology of Law, 7*: 37–60.

Carson, W.-G. (1980). The institutionalization of ambiguity: Early British factory acts. In G. Geis and E. Stotland (Eds.), *White collar crime: Theory and research*. London, UK: Sage.

Carson, W.-G. (1982). *The other price of Britain's oil*. London, UK: Martin Robertson.

Carson, W. (1985). Hostages to history: Some aspects of the occupational health and safety debate in historical perspective. In W. Creighton and N. Gunningham (Eds.), *The industrial relations of health and safety*. London, UK: Croom Helm.

Carson, W.-G., and Johnstone, R. S. (1990). The dupes of hazard: Occupational health and safety and the Victorian sanctions debate. *Australian and New Zealand Journal of Sociology, 26*: 126–141.

Cassell, J. (c. 1880s [no date]). *Cassell's Household Guide. Vol. 1* (New and Rev. Ed., 4 Vol.). Retrieved July 8, 2013 from http://www.victorianlondon.org/cassells/cassells-23.htm

CBS News. (2010). *BP's disaster: No surprise to folks in the know.* Retrieved June 23 from http://www.cbsnews.com/2100-215_162-6605248.html

Chandola, T., Marmot, M., and Siegrist, J. (2007). Failed reciprocity in close social relationships and health: Findings from the Whitehall II study. *Journal of Psychosomatic Research, 63*: 403–411.

REFERENCES / 221

Chief Inspector of Factories and Shops. (1897). *Annual Report, 1897.* Brisbane, Australia: Queensland Government.

Christian Science Monitor. (2010). US rig inspectors received gifts from oil companies, report says. Retrieved May 25 from http://www.csmonitor.com/USA/2010/0525/US-rig-inspectors-received-gifts-from-oil-companies-report-says

Christian Science Monitor. (2011). *Majority of blame for Gulf oil spill lies with BP, two US agencies find.* Retrieved September, 14 from http://www.csmonitor.com/USA/2011/0914/Majority-of-blame-for-Gulf-oil-spill-lies-with-BP-two-US-agencies-find

Christiani, D. (2011). Combating the environmental causes of cancer. *New England Journal of Medicine, 364*: 2266–2268.

Clapp, R. W., Howe, G. K., and Jacobs, M. (2006). Environmental and occupational causes of cancer re-visited. *Journal of Public Health Policy, 27*: 61–76.

Clarke, M., Lewchuk, W., de Wolff, A., and King, A. (2007). "This just isn't sustainable": Precarious employment, stress and workers' health. *International Journal of Law and Psychiatry, 30*: 311–326.

Clauss, C. A., Berzon, M., and Bertin, J. (1993). Litigating reproductive and developmental health in the aftermath of UAW versus Johnson Controls. *Environmental Health Perspectives Supplements, 101*(Suppl. 2): 205–220.

Cloutier, E., Duguay, P., Vézina, S., and Prud'homme, P. (2011). Accidents du travail. In M. Vézina, E. Cloutier, S. Stock, K. Lippel, É. Fortin, A. Delisle, M. St-Vincent, A. Funes, P. Duguay, S. Vézina, and P. Prud'homme (Eds.), *Enquête québécoise sur des conditions de travail, d'emploi, et de santé et de sécurité du travail (EQCOTESST)* (pp. 531-590). Montréal, Canada: Institut de recherche Robert Sauvé en santé et sécurité du travail, Institut national de santé publique du Québec, Institut de la Statistique du Québec.

Cloutier, E., Lippel, K., Bouliane, N., and Boivin, J.-F. (2011). Description des conditions de travail et d'emploi au Québec. In M. Vézina, E. Cloutier, S. Stock, K. Lippel, É. Fortin, A. Delisle, M. St-Vincent, A. Funes, P. Duguay, S. Vézina, and P. Prud'homme (Eds.), *Enquête québécoise sur des conditions de travail, d'emploi, et de santé et de sécurité du travail (EQCOTESST)* (pp. 59-158). Montréal, Canada: Institut de recherche Robert Sauvé en santé et sécurité du travail, Institut national de santé publique du Québec, Institut de la Statistique du Québec.

Commission de la Santé et de la Sécurité au Travail. (2011). *La modernisation du régime de santé et sécurité du travail.* Québec, Canada: Commission de la santé et de la sécurité au travail.

Confederation of British Industry (CBI). (2009). *CBI response to the consultation on the health and safety executive's strategy.* London, UK.

Confederation of British Industry (CBI). (2012). Retrieved February 22, 2012 from http://www.cbi.org.uk/search/?search=occupational+cancers

Corgey, D. (2010). Lack of union workers hurts offshore oil industry. *Houston Chronicle,* Opinion. Retrieved June 10 from http://www.chron.com/opinion/outlook/article/Lack-of-union-workers-hurts-offshore-oil-industry-1716136.php

Cox, R., and Lippel, K. (2008). Falling through the legal cracks: The pitfalls of using workers' compensation data as indicators of work-related injuries and illnesses. *Policy and Practice in Health and Safety, 6*(2): 9–30.

Cranford, C., and Vosko, L. (2006). Conceptualising precarious employment: Mapping the work across social location and occupational context. In L. Vosko (Ed.), *Precarious employment: Understanding labour market insecurity in Canada* (pp. 43–66). Montreal, Canada: McGill-Queens University Press.

Croidieu, S., Charbotel, B., Vohito, M., Renaud, L., Jaussaud, J., Bourboul, C., Ardiet, D., Imbard, I., Guerin, A., and Bergeret, A. (2008). Call-handlers' working conditions and their subjective experience of work—A transversal study. *International Archive of Occupational and Environmental Health, 82*: 67–77.

Cullen, W. D. (1990). *The public inquiry into the Piper Alpha disaster* (Vols. 1, 2). London, UK: Stationery Office.

Cutler, T., and James, P. (1996). Does safety pay? A critical account of the Health and Safety executive document: The Costs of Accidents. *Employment & Society, 10*(4): 755–765.

Dalton, A. (1979). *Asbestos: Killer dust*. London, UK: BSSRS.

David, H., Cloutier, E., and Ledoux, E. (2011). Precariousness, work organization and occupational health: The case of nurses providing home care services in Québec. *Policy and Practice in Health and Safety, 9*(2): 27–46.

Davidov, G. (2010). The enforcement crisis in labour law and the fallacy of voluntarist solutions. *International Journal of Comparative Labour Law and Industrial Relations, 26*: 61–81.

Davies, J. S. (2011). *Challenging governance theory: From networks to hegemony*. Bristol, UK: Policy Press.

Davies, J., Kemp, G., and Frostick, S. (2007). *An investigation of reporting of workplace accidents under RIDDOR using the Merseyside Accident Information Model. Research Report RR528*. Norwich, UK: HSE Books.

Davis, C. (2004). *Making companies safe: What works?* London, UK: Centre for Corporate Accountability.

Dawson, S., Willman, P., Bamford, M., and Clinton, A. (1988). *Safety at work: The limits of self-regulation*. Cambridge, UK: Cambridge University Press.

Dean Report. (2010). Expert advisory panel on occupational health and safety. (2010). *Report and recommendations to the Minister of Labour*. Retrieved December from http://www.labour.gov.on.ca/english/hs/pdf/eap_report.pdf

Deepwater Horizon Study Group. (2011). *Investigation of the Macondo well blowout disaster, final report, University of California and Center for Catastrophic Risk Management*. Retrieved March 1 from http://ccrm.berkeley.edu/pdfs_papers/bea_pdfs/DHSG FinalReport-March2011-tag.pdf

Dembe, A. E. (1996). *Occupation and disease. How social factors affect the conception of work-related disorders*. New Haven, CT: Yale University Press.

Dembe, A., Erickson, J., Delbos, R., and Banks, S. (2005). The impact of overtime and long work hours on occupational injuries and illnesses. *Occupational and Environmental Medicine, 62*: 588–597.

Devine, C., Farrell, T., Blake, C., Jastran, M., Wethington, E., and Bisogni, C. (2009). Work conditions and the food choice coping strategies of employed parents. *Journal of Nutrition Education and Behavior, 41*(5): 365–370.

Dew, K. (2011). Pressure to work through periods of short term sickness can have long term negative effects on health and productivity. *British Medical Journal, 341*: 1–2.

Doll, R., and Peto, R. (1981). The causes of cancer: Quantitative estimates of avoidable risks of cancer in the United States today. *Journal of National Cancer Institute, 66*: 1191–1308.

Dorman, P. (2000). If safety pays, why don't employers invest in it? In K. Frick, P. L. Jensen, M. Quinlan, and T. Wilthagen (Eds.), *Systematic occupational health and safety management*. Oxford, UK: Pergamon Press.

Doyal, L., Epstein, S., Green, K., et al. (1983). *Cancer in Britain: The politics of prevention*. London, UK: Pluto Press.

REFERENCES / 223

Doyal, L., with Pennell, I. (1979). *The political economy of health*. London, UK: Pluto Press.

Du & Jobbet. (2011). Nyheter; Här står striderna hetast. *Du & Jobbet, 4*: 55–56.

Du & Jobbet. (2012). Nyheter; Du & Jobbets studie: 96 000 skyddsombud. *Du & Jobbet, 2*: 45–46.

Dubowski, S. (2010, November 1). Complexity, confusion stymies C-45 charges. *Canadian Occupational Health and Safety*. Retrieved August 8, 2011 from http://www.cos-mag.com/Legal/Legal-Stories/Complexity-confusion-stymies-C-45-charges.html

Dukes, R. (2008). Constitutionalizing employment relations: Sinzheimer, Kahn-Freund, and the role of labour Law. *Journal of Law and Society, 35*: 341–366.

Dukes, R. (2009). Otto Kahn-Freund and collective laissez-faire: An edifice without a keystone? *Modern Law Review, 72*: 220–246.

Dumais, L., Messing, K., Seifert, A. M., Courville, J., and Vézina, N. (1993). Make me a cake as fast as you can: Determinants of inertia and change in the sexual division of labour of an industrial bakery. *Work, Employment & Society, 7*(3): 363–382.

DWP. (2011a, November). The government response to the Löfstedt Report. *Department of Work and Pensions*. Retrieved May 20, 2012 from www.dwp.gov.uk/docs/ lofstedt-report-respnse.pdf

DWP. (2011b). *Good health and safety, good for everyone. The next steps in the government's plans for reform of the health and safety system in Britain*. Retrieved May 2012 from http://www.dwp.gov.uk/docs/good-health-and-safety.pdf

Dwyer, T. (1991). *Life and death at work: Industrial accidents as a case of socially produced error*. New York, NY: Plenum Press.

Dyreborg, J. (2011). Safety matters have become too important for management to leave it up to the workers—The Nordic OSH model between implicit and explicit frameworks. *Nordic Journal of Working Life Studies, 1*(1): 135–160.

Eakin, J. (1992). Leaving it up to the workers: Sociological perspective on the management of health and safety in small workplaces. *International Journal of Health Services, 22*: 689–704.

Edwards, C. (2005). Where are all the C-45 OHS prosecutions? [Draft article]. Retrieved August 8, 2011 from http://www.heenanblaikie.com/en/media/pdfs/pdf/Where_Are_All_the_Bill_C-45_Prosecutions.pdf;jsessionid=94884D7F485727D18FFBDE0ED30ABFCA

Edwards, P. (2013). Project manager of Metron Highrise tragedy still faces trial. *Toronto Star* (10 January 2013).

Ehrenreich, B., and Hochschild, A. R. (2003). Introduction. In B. Ehrenreich and A. R. Hochschild (Eds.), *Global woman: Nannies, maids, and sex workers in the new economy*. New York, NY: Henry Holt Press.

Eilperin, J. (2010). Seeking answers in the Minerals Management Service's flawed culture. *The Washington Post*, Retrieved August 24 from http://www.washingtonpost.com/wp-dyn/content/article/2010/08/24/AR2010082406771.html?sid=ST2011010305686

Eliason, P. (2008). *Toxics use reduction and cancer prevention*. Stirling University (2008) International Conference on Occupational and Environmental Carcinogen. Papers at http://www.nmh.stir.ac.uk/research/occupational-environmental-cancer.php

Elovainio, M., Ferrie, J., Singh-Manoux, A., Gimeno, D., De Vogli, R., Shipley, M., et al. (2009). Cumulative exposure to high-strain and active jobs as predictors of cognitive function: The Whitehall II study. *Occupational and Environmental Medicine, 66*: 32-37.

Elovainio, M., Kivimaki, M., Linna, A., Brockner, J., van den Bos, K., Greenberg, J., Pentti, J., Virtanen, M., and Vahtera, J. (2010). Does organizational justice protect from sickness absence following a major life event? A Finnish public sector study. *Journal of Epidemiology and Community Health, 64*: 470–472.

224 / SAFETY OR PROFIT?

Engdahl, W. (2004). *A century of war. Anglo-American oil politics and the New World Order.* London, UK: Pluto Press.

Epstein, S. (1978). *The politics of cancer.* San Francisco, CA: Sierra Club Books.

Estlund, C. (2010). *Regoverning the workplace.* New Haven, CT: Yale University Press.

Eurofound (European Foundation for the Improvement of Living and Working Conditions). (2008). Femmes au travail: Les voies de l'égalité. *Note d'information,* 1–14.

Eurofound (European Foundation for the Improvement of Living and Working Conditions). (2011). *Working time developments.* Dublin, Ireland: European Foundation.

Euronews. (2012). Italy court finds company owners guilty in asbestos case. Retrieved February 23, 2012 from http://www.euronews.net/2012/02/13/italy-court-finds-company-owners-guilty-in-asbestos-case/

European Agency for Safety and Health at Work. (2003). *Gender issues in safety and health at work—A review.* Luxembourg: European Agency for Safety and Health at Work.

European Agency for Safety and Health at Work. (2007). *Expert forecast on emerging psychosocial risks related to occupational safety and health.* European Risk Observatory Report (p. 126). Luxembourg: European Agency for Safety and Health at Work.

European Agency for Safety and Health at Work. (2009). New and emerging risks in occupational safety and health. In E. R. Observatory (Ed.), *Outlook 1—New and emerging risks in occupational safety and health* (p. 28). Luxembourg: European Agency for Safety and Health at Work.

European Commission (EC). (2005). *Communication from the Commission on an EU common methodology for assessing administrative costs imposed by legislation,* COM (2005) 518 Final. 21.10.2005.

European Commission (EC). (2006). *Measuring administrative costs and reducing administrative burdens in the European Union,* COM (2006) 689 final and 690 final 14 November 2006.

European Commission (EC). (2007). *Commission Decision of 31 August 2007 setting up the High Level Group of Independent Stakeholders on Administrative Burdens,* Brussels, 2007.

European Commission (EC). (2009). *Reducing Administrative Burdens in the European Union, Annex to the 3rd Strategic Review on Better Regulation,* Brussels 28.1.2009 COM (2009) 16 final.

European Public Health Association (EPHA). (2012). Unhealthy influence: Patients groups and industry funding. Retrieved February 22, 2012 from http://www.epha.org/IMG/pdf/Unhealthy_influence_final.pdf

Expert Advisory Panel on Occupational Health and Safety. (2010a). *Report and recommendations to the Minister of Labour* (Dean Report). Canadian Government. Retrieved August 8, 2011 from http://www.labour.gov.on.ca/english/hs/pdf/eap_report.pdf

Expert Advisory Panel on Occupational Health and Safety. (2010b). *An update from the chair.* June 2010.

Expert Advisory Panel on Occupational Health and Safety. (2010c). *Report and recommendations to the Minster of Labour.* Retrieved December 2010 from http://www.labour.gov.on.ca/english/hs/pdf/eap_report.pdf

Federation of Small Businesses (FSB). (2012). *Health matters: The small business perspective.* London, UK: FSB. Retrieved December 2012 from http://www.fsb.org.uk/policy images/fsb%20health%20matters%20report.pdf

Ferrao, V., and Williams, C. (2011). *Women in Canada: A gender-based statistical report.* Statistics Canada: catalogue #89-503-XIE.

REFERENCES / 225

Ferrie, J., Martinkainen, P., Shipley, M., Marmot, M., Stansfeld, S., and Smith, G. (2001). Employment status and health after privatization in white collar civil servants: Prospective cohort study. *British Medical Journal, 322*: 1–7.

Ferrie, J., Shipley, M., Marmot, M., Stansfeld, S., and Smith, G. (1998). An uncertain future: The health effects of threats to employment security in white collar men and women. *American Journal of Public Health, 88*(7): 1030–1036.

Fidler, R. (1985). The Occupational Health and Safety Act and the Internal Responsibility System. *Osgoode Hall Law Journal, 24*: 315–352.

Fine, J. (2006). *Worker centers: Organizing communities at the edge of the dream.* Ithaca, NY: Cornell University Press.

Fine, J., and Gordon, J. (2010). Strengthening labor standards enforcement through partnership with workers organizations. *Politics & Society, 38*: 552–585.

Foley, M., Silverstein, B., Polissar, N., and Neradilek, B. (2009). Impact of implementing the Washington State Ergonomics Rule on employer reported risk factors and hazard reduction activity. *American Journal of Industrial Medicine, 52*: 1–16.

Fooks, G., Bergman D., and Rigby, B. (2007). *International comparison of (a) techniques used by state bodies to obtain compliance with health and safety law and accountability for administrative and criminal offences and (b) sentences for criminal offences.* RR607, Norwich, UK: HSE Books.

Frick, K. (1994). *Från sidovagn till integrerat arbetsmiljöarbete—Arbetsmiljöstyrning som ett ledningsproblem i svensk industri.* Stockholm, Sweden: Föreningen för Arbetarskydd.

Frick, K. (2004). Too much ambivalence in Australian OHS policies? *Journal of Occupational Health and Safety Australia-New Zealand, 20*(October): 395–400.

Frick, K. (2010). The Semi-Mandatory Implementation of EU´s Work Related Stress Agreement (WRS) in Sweden. *National Report to the EU-study on the Implementation of the European Social Partners' Agreement on Work Related Stress.* Västerås, Sweden: Mälardalen University.

Frick, K. (2011b). Implementing systematic work environment management in Sweden— Interpretation by SWEA and supervision by its labour inspectors. In D. Walters, R. Johnstone, K. Frick, M. Quinlan, G. Baril-Gingras, and A. Thebaud-Mony (Eds.), *Regulating workplace risk—A comparative study of inspection regimes in times of change.* Cheltenham, UK: Edward Elgar.

Frick, K., Eriksson, O., and Westerholm, P. (2005). Work environment policy and the actors involved. In R. Gustafsson and I. Lundberg (Eds.), *Worklife and health in Sweden 2004.* Stockholm, Sweden: National Institute for Working Life.

Frick, K., Eriksson, S., and Westerholm, P. (2009). Health and safety representation in small firms: A Swedish success under threat. In D. Walters and T. Nichols (Eds.), *Workplace health and safety—International perspective on worker representation.* Basingstoke, UK: Palgrave Macmillan.

Frick, K., Eriksson, S., and Westerholm, P. (2011a). The regulation of systematic work environment management in Sweden—Higher ambitions in a weaker Swedish work environment system. In D. Walters, R. Johnstone, K. Frick, M. Quinlan, G. Baril-Gingras, and A. Thebaud-Mony (Eds.), *Regulating workplace risk—A comparative study of inspection regimes in times of change.* Cheltenham, UK: Edward Elgar.

Frick, K., and Forsberg, A. (2010). *Hälsosam-projektet—Framgångar och svårigheter i att genomföra FAS 05 för bättre samverkan och arbetsmiljö i Leksand.* Falun, Sweden: Dalarnas Forskningsråd.

226 / SAFETY OR PROFIT?

Frick, K., Jensen, P. L., Quinlan, M., and Wilthagen, T. (2000). Systematic occupational health and safety management—An introduction to a new strategy for occupational health and safety well-being. In K. Frick, P. L. Jensen, M. Quinlan, and T. Wilthagen (Eds.), *Systematic occupational health and safety management*. Oxford, UK: Pergamon Press.

Fristedt, K. (2011). Direct information by mail from Karin Fristedt, OHS officer at SACO, 2011-09-09.

Fritschi, L., and Driscoll, T. (2006). Cancer due to occupation in Australia. *Australian and New Zealand Journal of Public Health, 30*: 213–219.

Fudge, J. (2011). Global care chains, employment agencies, and the conundrum of jurisdiction: Decent work for domestic workers in Canada. *Canadian Journal of Women and the Law, 23*(1): 235–264.

Fuller, S., and Vosko, L. F. (2008). Temporary employment and social inequality in Canada: Exploring intersections of gender, race and immigration status. *Social Indicators Research, 88*: 31–50.

Gash, V., Mertens, A., and Romeau Gordo, L. (2006). Are fixed-term jobs bad for your health? A comparison of West Germany and Spain. *IAB Discussion Paper No. 8/2006*, Berlin, Germany.

Geldart, S., Shannon, H. S., and Lohfeld, L. (2005). Have companies improved their health and safety approaches over the last decade? A longitudinal study. *American Journal of Industrial Medicine, 47*: 227–236.

Geldart, S., Smith, C. A., Shannon, H. S., and Lohfeld, L. (2010). Organizational practices and workplace health and safety: A cross-sectional study in manufacturing companies. *Safety Science, 48*(5): 562–569.

Gellerstedt, S. (2007). *Samverkan för bättre arbetsmiljö—Skyddsombudens arbete och erfarenheter*. Stockholm, Sweden: Landsorganizationen i Sverige.

Gellerstedt, S. (2010). *Telephone interview with Sten Gellerstedt, LO officer on safety representative issues,* December 2, 2011.

Gellerstedt, S. (2011). *Fler arbetare måste få utvecklande jobb—Inte digital Taylorism.* Stockholm, Sweden: Landsorganizationen i Sverige.

Gellerstedt, S. (2012). Mail with a table of members and safety representatives per LO-union, February 27, 2012.

Gherardi, S., and Nicolini, D. (2000a). To transfer is to transform: The circulation of safety knowledge. *Organization, 7*(2): 329–348.

Gherardi, S., and Nicolini, D. (2000b). The organizational learning of safety in communities of practice. *Journal of Management Inquiry, 9*(1): 7–18.

Gill, D. A., Picou, J. S., and Ritchie, L. A. (2012). The Exxon Valdez and BP oil spills: A comparison of initial social and psychological impacts. *American Behavioral Scientist, 56*(1): 3–23.

Gillard, M., Jones, M., and Rowell, A. (2005). Northern exposure. In C. Woolfson and M. Beck (Eds.), *Corporate social responsibility failures in the oil industry* (pp. 163–76). Amityville, NY: Baywood.

Glasbeek, H. (1998). Occupational health and safety law: Criminal law as a political tool. *Australian Journal of Labour Law, 11*: 95.

Glasbeek, H. (2008). Book review: Varieties of capitalism, corporate governance and employees. *Australian Journal of Corporate Law, 22*: 293–304.

Grant, K. R., Amaratunga, C., Armstrong, P., Boscoe, M., Pederson, A., and Wilson, K. (2004). *Caring for/caring about. Women, home care and unpaid caregiving.* Aurora, Canada: Garamond Press.

REFERENCES / 227

Gray, C. G. (2002). A socio-legal ethnography of the right to refuse dangerous work. *Studies in Law, Politics and Society, 24*: 133–169.

Gray, G. C. (2009). The responsibilization strategy of health and safety. *British Journal of Criminology, 49*: 326–342.

Greenberg, M. (2005). Three tears for EMAS. *Occupational Medicine, 55*: 73.

Greenfieldboyce, N. (Writer). (2012). New silica rules languish in regulatory black hole [Radio]. In N. P. Radio (Producer). Retrieved February 20, 2012 from http://www.npr.org/2012/02/01/146168033/new-silica-rules-languish-in-regulatory-black-hole

Grusenmeyer, C. (2007). Sous-traitance et accidents. Exploitation de la base de données EPICEA *NS 226* (p. 124). Retrieved February 23, 2012 from http://www.inrs.fr/inrs-pub/inrs01.nsf/IntranetObject-accesParReference/NS%20266/$FILE/ns266.pdf

Guardian. (2010). *Barack Obama finding out "whose ass to kick" over oil spill.* Retrieved June 8 from http://www.guardian.co.uk/world/2010/jun/08/barack-obama-kick-ass-gulf

Gunningham, N. (2007). *Mine safety: Law, regulation and policy.* Sydney, Australia: Federation Press.

Gunningham, N. (2008). Occupational health and safety, worker participation and the mining industry in a changing world of work. *Economic and Industrial Democracy, 29*: 336–361.

Gunningham, N., and Johnstone, R. (1999). *Regulating workplace safety: Systems and sanctions.* Oxford, UK: Oxford University Press.

Gunningham, N., Johnstone, R., and Rozen, P. (1996). *Enforcement measures for occupational health and safety in New South Wales: Issues and options.* Sydney, Australia: WorkCover Authority of New South Wales.

Gunningham, N., Sinclair, D., and Burritt, P. (1998). *On-the-spot fines and the prevention of injury and disease—The experience of Australian workplaces.* Sydney, Australia: National Occupational Health and Safety Commission.

Habib, R., Fathallah, F. A., and Messing, K. (2010). Similarities and differences between household work and tasks in selected comparable occupations. *International Journal of Occupational Safety and Ergonomics, 16*(1): 113–128.

Hall, A., Forrest, A., Sears, A., and Carlan, N. (2006). Making a difference: Knowledge activism and worker representation in joint OHS committees. *Relations Industrielles/Industrial Relations, 61*: 408–436.

Hall, A., and Johnstone, R. (2005). Exploring the recriminalising of OHS breaches in the context of industrial death. *The Flinders Journal of Law Reform, 8*(1): 57–92.

Haines, F., and Hall, A. (2004). The law and order debate in occupational health and safety. *Journal of Occupational Health and Safety—Australia and New Zealand, 19*: 263.

Ham, J. (1976). *Report of the Royal Commission on the Health and Safety of Workers in Mines.* Ministry of the Attorney General, Province of Ontario, Canada.

Hämäläinen, P., Takala, J., and Saarela, K. L. (2005). *Global estimates of fatal work-related diseases and occupational accidents.* Amsterdam, The Netherlands: Elsevier.

Hämäläinen, P., Takala, J., and Saarela, K. L. (2007). Global estimates of fatal work-related diseases. *American Journal of Industrial Medicine, 50*: 28–41.

Hampton, P. (2005). *Reducing administrative burdens: Effective inspection and enforcement.* London, UK: HM Treasury/HMSO.

Hanley, J., Premji, S., Messing, K., and Lippel, K. (2010). Action research for the health and safety of domestic workers in Montreal: Using numbers to tell stories and effect change. *New Solutions, 20*(4): 421–439.

228 / SAFETY OR PROFIT?

Hansen, J., and Stevens, R. G. (2011). Case-control study of shift-work and breast cancer risk in Danish nurses: Impact of shift systems. *European Journal of Cancer*, August 16. [Epub ahead of print].

Harish, C. et al. (2010). Effectiveness of Canada's employment equity legislation for women (1997–2004): Implications for policy makers. *Relations Industrielles, 65*: 304–329.

Harremoës, P., Gee, D., and MacGarvin, M. (2001). *Late lessons from early warnings: The precautionary principle 1896–2000*. Brussels, Belgium: European Environment Agency, European Commission.

Harrington, M., and Gill, F. (1983). *Occupational health*. Oxford, UK: Blackwell Scientific Publications.

Harrison, B. (1996). *Not only the "dangerous trades": Women's work and health in Britain, 1880–1914*. London, UK: Taylor and Francis.

Hart, S. M. (2010). Self-regulation, corporate social responsibility and the business case: Do they work in achieving workplace equality and safety? *Journal of Business Ethics, 92*: 585–600.

Hasle, P. (2007). Outsourcing and employer responsibility: A case study of occupational health and safety in the Danish public transport sector. *Relations Industrielles, 62*(1): 96–117.

Hatton, E. (2011). *The temp economy*. Philadelphia, PA: Temple University Press.

Hazards. (2009). *Zero cancer campaign*. Retrieved February 22, 2012 from http://www.hazards.org/cancer/preventionkit/

Hazards. (n.d.). *Work cancer prevention kit*. Available at http://www.hazards.org/cancer/preventionkit/part1.htm

Health and Safety Executive (HSE). (1997). *Successful health and safety management*. HS(G)65 (rev. ed.). Sudbury, UK: HSE Books.

Health and Safety Executive (HSE). (2004). *HSE launches new business benefits of health and safety web pages*. HSE Press Release: E137:04. Retrieved October 4, 2004 from http://www.hse.gov.uk/press/2004/e04137.htm

Health and Safety Executive (HSE). (2006). *Managing shift work: Health and safety guidance*. Norwich, UK: HSE Books.

Health and Safety Executive (HSE). (2007). Key Programme 3 Asset Integrity Programme: A report by the Offshore Division of HSE's Hazardous Installations Directorate. *HSE*. Available at http://www.hse.gov.uk/offshore/kp3.pdf

Health and Safety Executive (HSE). (2009). *HSE out of hours contact web page*. Retrieved May 9, 2009 from http://www.hse.gov.uk/contact/outofhours.htm

Health and Safety Executive (HSE). (2010a). *European comparisons*. Retrieved January 8, 2011 from http://www.hse.gov.uk/statistics/european/fatal.htm

Health and Safety Executive (HSE). (2010b). *HSE welcomes Lord Young's report on health and safety*. Available at http://www.hse.gov.uk/press/2010/hse-lordyoungreport.htm

Health and Safety Executive (HSE). (2011a). *The health and safety executive statistics 2010/11*. Available at http://www.hse.gov.uk/statistics/overall/hssh1011.pdf

Health and Safety Executive (HSE). (2011b). *Making health & safety pay*. Retrieved April 5, 2011 from http://www.hse.gov.uk/betterbusiness/large/hsebooklet.pdf

Health and Safety Executive (HSE). (2013). The working time regulations 1998 and amended in 2013. Retrieved July 5, 2013 from http://www.hse.gov.uk/contact/tag/workingtime.directive.htm

Health and Safety Executive (HSE). (n.d.). *About corporate responsibility*. Available at http://www.hse.gov.uk/corporateresponsibility/about.htm

REFERENCES / 229

Heinrich, J. (2012). Pink Ribbon Inc. *The Gazette*, Montreal. January 30, 2012. Retrieved February 22, 2012 from http://www.montrealgazette.com/health/Pink+Ribbons+fine+line+between+fundraising+feminism/6063744/story.html?id=6063744

Hirschman, A. (1970). *Exit, voice and loyalty. Responses to decline in firms, organizations and states.* Cambridge, MA: Harvard University Press.

Hirsh, C. E. (2009). The strength of weak enforcement: The impact of discrimination charges, legal environments, and organizational conditions on workplace segregation. *American Sociological Review, 74*: 245–271.

Hochschild, A. R. (2003). *The managed heart: Commercialization of human feeling.* Berkeley and Los Angeles: University of California Press.

Home Office. (1933). *Annual Report of the Chief Inspector of Factories and Workshops for the Year 1932: 9* Cmnd 4377. London, UK: Stationary Office.

Homkes, R. (2011). Enhancing management quality. *Centrepiece, 15*(3) Winter: 2–6.

Hopkins, A. (2011). Management walk-arounds: Lessons from the Gulf of Mexico oil well blowout, National Research Centre for Occupational Health and Safety Regulation. *Regnet*, Australian National University, Working Paper 79, February. Available at http://www.aimwa.com/Pages/Open%20Programs/~/media/Files/PDF/Open%20Programs/WP%2079%20Hopkins%20Gulf%20of%20Mexico.ashx

Hovden, J., Lie, T., Karlsen, J., and Alteren, B. (2008). The safety representative under pressure: A study of occupational health and safety management in the Norwegian oil and gas industry. *Safety Science, 46*: 493–509.

HRSDC. (2011). *Labour Market Bulletin (Fall 2011)*, Canadian Government. Retrieved August 8, 2011 from http://www.hrsdc.gc.ca/eng/workplaceskills/labour_market_information/bulletins/on/on-lmb-2011fall.pdf

Hueper, W. (1942). *Occupational tumours and allied diseases.* Springfield, IL: Charles C. Thomas.

Hughes, P. (2007). *Tackling "rogues."* Health and Safety Executive Board Paper HSE/07/122: December 5.

Hyman, R. (2010). Editorial. *European Journal of Industrial Relations, 16*(4).

Iavicoli, S., Natali, E., Deitinger, P., Maria Rondinone, B., Ertel, M., Jain, A., and Leka, S. (2011). Occupational health and safety policy and psychosocial risks in Europe: The role of stakeholders' perceptions. *Health Policy, 101*(1): 87–94.

IF Metall. (2011). Interview January 28, 2011 with Stefan Wiberg and Bo Andersson, OHS officers. Stockholm, Sweden: IF Metall.

Imander, B. (1999). *Skyddsombudets värld—Fallstudier om skyddsombudets uppfattning om sin roll och sitt uppdrag.* Uppsala, Sweden: Pedgogiska institutionen, Uppsala universitet.

Institut de la Statistique du Québec. (2008a). *Répartition des femmes de 25-54 ans, personne de référence ou conjointe, avec ou sans enfants de moins de 25 ans à la maison selon la situation familiale, l'occupation d'un emploi et le régime de travail, Québec, 1997 et 2008.* Retrieved February 28, 2012 from http://www.stat.gouv.qc.ca/donstat/societe/famls_mengs_niv_vie/tendances_travail/tab_web_fam_tab_1.htm

Institut de la Statistique du Québec. (2008b). *Répartition des hommes de 25-54 ans, personne de référence ou conjoint, avec ou sans enfants de moins de 25 ans à la maison selon la situation familiale, l'occupation d'un emploi et le régime de travail, Québec, 1997 et 2008.* Retrieved February 28, 2012 from http://www.stat.gouv.qc.ca/donstat/societe/famls_mengs_niv_vie/tendances_travail/tab_web_fam_tab_4.htm

Institut de la Statistique du Québec. (2010). Les 25 principales professions les plus fréquentes chez les hommes en 2006 selon le rang en 1991. In *Tendances Sociales du travail*:

230 / SAFETY OR PROFIT?

Institut de la statistique du Québec. Retrieved July 5, 2013 from http://www.stat.gouv. qc.ca/donstat/societe/famls_mengs_niv_vie/tendances_travail/tab3_p9106_ph.htm

Institute of Directors and Health and Safety Executive. (2009). *Leading health and safety at work. Leadership actions for directors and board members.* INDG417, 09/09, Sudbury, UK: HSE Books.

International Agency for Research on Cancer (IARC). (2011). The Asturius Declaration. Lyons: IARC. Retrieved December 4, 2012 from http://www.iarc.tr/en/media-centre/iarcnews/2011/asturiusdeclaration.php

International Labour Organization. (2010). *List of occupational disease (rev. 2010). Identification and recognition of occupational diseases: Criteria for incorporating in the ILO list of occupational diseases.* Geneva, Switzerland: ILO.

Ireland, P. (2005). Shareholder primacy and the distribution of wealth. *Modern Law Review, 68*: 49–81.

Iriat, J., de Oliveira, R., Xavier, S., da Silva Costa, A., de Araujo, G., and Santana, V. (2008). Representations of informal jobs and health risks among housemaids and construction workers. *Ciencia & Saude Coletivia, 13*(1): 164-174.

ITF (International Transportworkers' Federation) Seafarers. (2002). *Offshore workers in the spotlight.* Available at http://www.itfseafarers.org/offshore-spotlight.cfm

James, P., Johnstone, R., Quinlan, M., and Walters, D. (2007). Regulating supply chains for safety and health. *Industrial Law Journal, 36*(2): 163–187.

Järvholm, B. (2010). *Arbetsrelaterade dödsfall i Sverige. Kunskapsoversikt. Rapport 2010:3.* Stockholm, Sweden: Arbetsmiljöverket.

Jasanoff, S. (1987). Contested boundaries in policy-relevant science. *Social Studies of Science, 17*(2): 195–230.

Johansson, J. (2010). Telephone interview with Jan Johansson, OHS officer at Almega, December 8, 2010.

Johansson, L-B. (2010). Telephone interview with Lill-Britt Johansson, Kommunal, December 7, 2010.

Johansson, S. (2011). *Implementering av Lean produktion—En utmaning för fackföreningsrörelsen.* Working Paper. Luleå, Sweden: Department of Work Science, Luleå Technical University.

Johnson, B. (2010). *Kampen om sjukfrånvaron.* Lund, Sweden: Arkiv förlag.

Johnson, D. (2005). *What ever happened to the business case for safety?* ISHN E-Zine 4:9 (September 21, 2005). Retrieved April 11, 2011 from http://www.ishn.com/Articles/Newsletter_Archive/d5b62bf8f644c010VgnVCM100000f932a8c0

Johnson, J. (2008). Globalisation, workers' power and the psychosocial work environment— Is the demand-control-support model still useful in the neoliberal era? *Scandinavian Journal of Work Environment and Health,* (Suppl. 6): 15–21.

Johnstone, R. (2000). Occupational health and safety prosecutions in Victoria: An historical study. *Australian Journal of Labour Law, 13*: 113–142.

Johnstone, R. (2003). *Occupational health and safety, courts and crime: The legal construction of occupational health and safety offences in Victoria.* Sydney, Australia: Federation Press.

Johnstone, R. (2004a). *Occupational health and safety law and policy.* Sydney, Australia: Thomson Lawbook.

Johnstone, R. (2004b). Rethinking OHS enforcement. In E. Bluff, N. Gunningham, and R. Johnstone (Eds.), *OHS regulation for a changing world of work* (pp. 146–178). Sydney, Australia: Federation Press.

REFERENCES / 231

Johnstone, R. (2007). Are occupational health and safety crimes hostage to history?: An Australian perspective. In A. Brannigan and G. Pavlich (Eds.), *Governance and regulation in social life* (pp. 33–54). London, UK: Cavendish-Routledge.

Johnstone, R., and King, M. (2008). A responsive sanction to promote systematic compliance?: Enforceable undertakings in occupational health and safety regulation. *Australian Journal of Labour Law, 21*: 280–315.

Johnstone, R. (2009). From catch up to innovation: A history of occupational health and safety regulation in Queensland. In B. Bowden, S. Blackwood, C. Rafferty, and C. Allan (Eds.), *Work & strife in paradise: The history of labour relations in Queensland 1859–2009* (pp. 95–112). Sydney, Australia: Federation Press.

Johnstone, R., and Parker, C. (2010). *Enforceable undertakings in action—Report of a roundtable discussion with Australian regulators.* National Research Centre for Occupational Health and Safety Regulation Working Paper No. 71. Canberra, Australia: Regulatory Institutions Network, Australian National University.

Johnstone, R., Quinlan, M., and Mayhew, C. (2001). Outsourcing risk? The regulation of OHS where contractors are employed. *Comparative Labor Law and Policy Journal, 22*(2/3): 351–393.

Johnstone, R., Quinlan, M., and McNamara, M. (2011). OHS inspectors and psychosocial risk factors: Evidence from Australia. *Safety Science, 49*(4): 547–558.

Johnstone, R., Quinlan, M., and Walters, D. (2005). Statutory occupational health and safety workplace arrangements for the modern labour market. *Journal of Industrial Relations, 47*(1): 93–116.

Jusek. (2011). Telephone and mail information from Emma Olseni Ehnmark, Kerstin Peangberg and Marie Thuresson Jusek, September 23, 2011 and October 3, 2011.

Kamienski, A. (1979). *Utbildning i bättre arbetsmiljö—Framsteg och problem.* Stockholm, Sweden: Arbetarskyddsfonden.

Karasek, R. (1979). Job demands, job decision latitude, and mental strain: Implications for job re-design. *Administrative Science Quarterly, 24*: 285–308.

Karasek, R., Baker, D., Marxer, F., Ahlbom, A., and Theorell, T. (1981). Job decision latitude, job demands, and cardiovascular disease: A prospective study of Swedish men. *American Journal of Public Health, 71*: 694–705.

Kauppinen, T., Toikkanen, J., Pedersen, D., et al. (2000). Occupational exposure to carcinogens in the European Union. *Occupational and Environmental Medicine, 57*: 10–18.

Keen, C., Coldwell, M., McNally, K., Baldwin, P., and McAlinden, J. (2010). *Occupational exposure to MbOCA and isocyanates in polyurethane manufacture.* HSE RR 828. Buxton, UK: HSL. Available at http://www.hse.gov.uk/research/rrpdf/rr828.pdf

Kellen, E., Zeegers, M. P., Hond, E. D., and Buntinx, F. (2007). Blood cadmium may be associated with bladder carcinogenesis: The Belgian case-control study on bladder cancer. *Cancer Detection and Prevention, 31*: 77–82. Epub February 12, 2007.

Keisu Lennerlöf, L. (1981). *Rollen som skyddsombud—Rapport från 257 skyddsombud inom SACO/SR.* Stockholm, Sweden: SACO/SR.

Keith, N. (2011). Regulators gone wild! *Canadian Occupational Health and Safety,* (January 27, 2011). Retrieved August 8, 2011 from http://www.cos-mag.com/Legal/Legal-Columns/regulators-gone-wild.html

Keith, N., and Abbott, A. (2011). *Acquittal in Quebec Bill C-45 Charges, Gowlings: Occupational Health and Safety Newsflash* (April 2011). Retrieved August 8, 2011 from http://www.gowlings.com/KnowledgeCentre/enewsletters/ohslaw/htmfiles/ohslaw 20110427.en.html

232 / SAFETY OR PROFIT?

Kim, I.-H., Khang, Y.-H., Muntaner, C., Chun, H., and Cho, S.-I. (2008). Gender, precarious work, and chronic diseases in South Korea. *American Journal of Industrial Medicine, 51*(10): 748–757.

Kim, M. (2011). Gender, work and resistance: South Korean textile industry in the 1970s. *Journal of Contemporary Asia, 41*(3): 411–430.

Kindenberg, U. (2011). Företagshälsovården i går, i dag, i morgon. In *SOU 2011: 63. Framgångsrik Företagshälsovård—Möjligheter och Metoder.* Stockholm, Sweden: Fritzes.

Kinnersly, P. (1973). *The hazards of work.* London, UK: Pluto Press.

Kivimaki, M., Honkonen, T., Wahlbeck, K., Elovainio, M., Pentti, J., Klaukka, T., Virtanen, M., and Vahtera, J. (2007). Organizational downsizing and increased use of psychotropic drugs among employees who remain in employment. *Journal of Epidemiology and Community Health, 61*: 154–158.

Kjaerheim, K. (1999). Occupational cancer research in the Nordic countries. *Environmental Health Perspectives, 107*(Suppl 2): 233–238.

Kjellberg, A. (2011). The decline in Swedish Union density since 2007. *Nordic Journal of Working Life Studies, 1*(1); 67–93.

Kome, P. (1998). *Wounded workers: The politics of musculoskeletal injuries.* Toronto, Canada: University of Toronto Press.

Kommunal. (2011). *Ständigt Standby—En Rapport om Visstidsanställdas Villkor.* Stockholm, Sweden: Kommunalarbetarförbundet.

Krawiec, K. D. (2003). Cosmetic compliance and the failure of negotiated governance. *Washington University Law Quarterly, 81*: 487–544.

Kronlund, J., Fleischauer, B., and Grünbaum, M. (1985). *Miljörörelse Men i Ordnade Former—Arbetsmiljölagen som Instrument för Miljöförändringar.* Stockholm, Sweden: Arbetslivscentrum.

Labbe, E., Moulin, J., Sass, C., Chatain, C., and Gerbaud, L. (2007). Relations entre formes particulieres d'emploi, vulnerabilite sociale et sante. *Archives des Maladies Professionnelles et de l'Environnement, 68*(4): 365–375.

Labonte, R., and Schrecker, T. (2007). Globalisation and social determinants of health: The role of the global marketplace. *Globalization and Health, 3*: 6.

Lacey, R., Lewis, M., and Sim, J. (2007). Piecework, musculoskeletal pain and the impact of workplace psychosocial factors. *Occupational Medicine, 57*: 430–437.

Landrigan, P. J., Espina, C., and Neira, M. (2011). Global prevention of environmental and occupational cancer. *Environmental Health Perspectives, 119*: a280–a281. Available at http://dx.doi.org/10.1289/ehp.1103871

Lärarnas Riksförbund. (2005). *Arbetsmiljöundersökning 2005, lärare.* Stockholm, Sweden: Lärarnas Riksförbund.

Levesque, C. (1995). State intervention in occupational health and safety: Labour-management committees revisited. In A. Giles and K. Wetzel (Eds.), *Proceedings of the XXXIst Conference of the Canadian Industrial Relations Association* (pp. 217–231). Toronto, Canada.

Levi, L. (2001). *Stress och Hälsa.* Stockholm, Sweden: Skandia.

Levinson, K. (2004). *Lokal Partssamverkan—En undersökning av Svenskt Medbestämmande.* Arbetsliv i omvandling 2004: 5. Stockholm, Sweden: Arbetslivsinstitutet

Lewchuk, W., Clarke, M., and de Wolff, A. (2008). Working without commitments: Precarious employment and health. *Work, Employment and Society, 22*(3): 387–406.

Lewchuk, W., Leslie, A., Robb, L. A., and Walters, V. (1996). The effectiveness of Bill 70 and Joint Health and Safety Committees in reducing injuries in the workplace: The case of Ontario. *Canadian Public Policy, 23*: 225–243.

REFERENCES / 233

Lillie, N. (2010). Bringing the offshore ashore: Transnational production, industrial relations and the reconfiguration of sovereignty. *International Studies Quarterly, 54*: 683–704.

Linhart, R. (1981). *The assembly line*. London, UK: John Calder.

Lippel, K. (1989). Workers' compensation and psychological stress claims in North American law: A microcosmic model of systemic discrimination. *International Journal of Law and Psychiatry, 12*(1): 41–70.

Lippel, K. (1998). Preventive reassignment of pregnant or breast-feeding workers: The Québec model. *New Solutions: A Journal of Environmental and Occupational Health Policy, 8*(2): 267–280.

Lippel, K. (1999). Workers' compensation and stress: Gender and access to compensation. *International Journal of Law and Psychiatry, 22*(5/6): 521–546.

Lippel, K. (2003). Compensation for musculo-skeletal disorders in Quebec: Systemic discrimination against women workers? *International Journal of Health Services, 33*(2): 253–281.

Lippel, K. (2005). Le harcèlement psychologique au travail : Portrait des recours juridiques au Québec et des décisions rendues par la Commission des lésions professionnelles. *Revue Pistes, 7*(3). Retrieved February 20, 2012 from http://www.pistes.uqam.ca/v7n3/articles/v7n3a13.htm

Lippel, K. (2006). Precarious employment and occupational health and safety regulation in Quebec. In L. Vosko (Ed.), *Precarious employment: Understanding labour market insecurity in Canada* (pp. 241–255). Montreal, Canada: McGill-Queens University Press.

Lippel, K. (2009). Le droit québécois et les troubles musculo-squelettiques: Règles relatives à l'indemnisation et à la prévention. *Revue Pistes, 11*(2): 1–14. Retrieved February 20, 2012 from http://www.pistes.uqam.ca/v11n2/articles/v11n2a3.htm

Lippel, K. (2011). Law, public policy and mental health in the workplace. *Healthcare Papers, 11*(Suppl.): 20–37.

Lippel, K., and Caron, J. (2004). L'ergonomie et la réglementation de la prévention des lésions professionnelles en Amérique du Nord. *Relations Industrielles/Industrial Relations, 58*(2): 233–270.

Lippel, K., MacEachen, E., Saunders, R., Werhun, N., Kosny, A., Mansfield, E., Carrasco, C., and Pugliese, D. (2011). Legal protections governing occupational health and safety and workers' compensation of temporary employment agency workers in Canada: Reflections on regulatory effectiveness. *Policy and Practice in Health and Safety, 9*(2): 69–90.

Lippel, K., Messing, K., Vézina, S., and Prud'homme, P. (2011). Conciliation travail et vie personnelle. In M. Vézina, E. Cloutier, S. Stock, K. Lippel, É. Fortin, A. Delisle, M. St-Vincent, A. Funes, P. Duguay, S. Vézina, and P. Prud'homme (Eds.), *Enquête Québécoise sur des Conditions de Travail, d'Emploi, et de Santé et de Sécurité du Travail (EQCOTESST)* (pp. 159–231). Montréal, Canada: Institut de recherche Robert Sauvé en santé et sécurité du travail, Institut national de santé publique du Québec, Institut de la Statistique du Québec.

Lippel, K., and Quinlan, M. (2011). Editorial: Regulation of psychosocial risk factors at work: An international overview. *Safety Science, 49*(4): 543–546.

Lippel, K., and Sikka, A. (2010). Access to workers' compensation benefits and other legal protections for work-related mental health problems: A Canadian overview. *Revue Canadienne de Santé Publique/Canadian Journal of Public Health, 101*(Suppl.1): S16–S22.

Lippel, K., Vézina, M., and Cox, R. (2011). Protection of workers' mental health in Québec: Do general duty clauses allow labour inspectors to do their job? *Safety Science, 49*: 582–590.

234 / SAFETY OR PROFIT?

Lippel, K., Vézina, M., Stock, S., and Funes, A. (2011). Violence au travail: Harcèlement psychologique, harcèlement sexuel et violence physique. In M. Vézina, E. Cloutier, S. Stock, K. Lippel, É. Fortin, A. Delisle, M. St-Vincent, A. Funes, P. Duguay, S. Vézina, and P. Prud'homme (Eds.), *Enquête Québécoise sur des Conditions de Travail, d'Emploi, et de Santé et de Sécurité du Travail (EQCOTESST)* (pp. 326–399). Montréal, Canada: Institut de recherche Robert Sauvé en santé et sécurité du travail; Institut national de santé publique du Québec; Institut de la Statistique du Québec.

Lipscomb, H., Kucera, K., Epling, C., and Dement, J. (2008). Upper extremity musculoskeletal symptoms and disorders among a cohort of women employed in poultry processing. *American Journal of Industrial Medicine, 51*: 21–36.

Livs. (2011). *Arbetsmiljöavtal inom Livsmedelsbranchen, Bilaga B till Livsmedelsavtalet.* Stockholm, Sweden: Livsmedelsarbetareförbundet.

Lloyd, C., and James, S. (2008). Too much pressure? Retailer power and occupational health and safety in the food processing industry *Work, Employment and Society, 22*(4): 713–730.

LO. (1997). *Hur fungerar arbetsmiljöarbetet?—En Enkätundersökning Bland LOs Skyddsombud.* Stockholm, Sweden: Landsorganizationen i Sverige.

LO. (2011). *Anställningsformer år 2011. Fast och Tidsbegränsat Anställda efter Klass och kön år 1990–2011.* Stockholm, Sweden: Landsorganization i Sverige.

Lobel, O. (2004). The renew deal: The fall of regulation and the rise of governance in contemporary legal thought, *Minnesota Law Review, 89*: 342–470.

Lobel, O. (2005). Interlocking regulatory and industrial relations: The governance of workplace safety. *Administrative Law Review, 57*: 1071–1152.

Löfstedt, R. E. (2011). *Reclaiming health and safety for all: An independent review of health and safety legislation.* London, UK: HMSO, November, Cmnd. 8219.

Loos, F., and Le Deaut, J. (2002). *Rapport Fait Au Nom de la Commission D'Enquete sur la Surete des Installations Industrielles et des Centres de Recherche et sur la Protection de Personnes et de L'environnment en cas D'Accident Industriel Majeur.* Assemblee Nationale No. 3559. Paris, France.

LO-tidningen. (2008a). Utbildning för skyddsombud stryps. *LO-tidningen,* November 11, 2008, p. 8.

LO-tidningen. (2008b). Regeringen slopade utbildningsstödet. *LO-tidningen,* November 11, 2008, p. 11.

LO-tidningen. (2012). En av tre nobbar LO-facken. *LO-tidningen,* February 10, 2012, p. 4.

Louie, A., Ostry, A., Keegal, T., Quinlan, M., Shoveller, J., and LaMontagne, A. (2006). Empirical study of employment arrangements and precariousness in Australia. *Relations Industrielles/Industrial Relations, 61*(3): 465–489.

Lukes, S. (1974). *Power—A radical view.* London, UK: Macmillan.

Lund, H. A. (2002). Integrerede ledelsessystemer og arbejdspladsdemokrati i et bæredygtighedsperspektiv. *Tidsskrift for Arbejdsliv, 4*(4): 39–58.

Lundh, C., and Gunnarsson, C. (1987). *Arbetsmiljö, arbetarskydd och utvärderingsforskning. Ett historiskt perspektiv.* Lund, Sweden: Wilber—Ekonomisk-historiska föreningen.

Lycke, A. (2011). Telephone interview with Alicia Lycke. OHS officer, Sveriges Läkarförbund, October 4, 2011.

Lysgaard, S. (1961). *Arbeiderkollektivet.* Oslo, Norway: Universitetsforlaget.

MacEachen, E., Ferrier, S., and Chambers, L. (2007). A deliberation on "hurt versus harm" logic in early-return-to-work policy. *Policy and Practice in Health and Safety, 5*(2): 41–62.

REFERENCES / 235

MacEachen, E., Kosny, A., Ferrier, S., Lippel, K., Neilson, C., Franche, R.-L., and Pugliese, D. (2011). The "ability" paradigm in vocational rehabilitation: Challenges in an Ontario injured worker retraining program. *Journal of Occupational Rehabilitation, 20*(4): 1–13.

Maconachie, G. (1986). *The origins, development and effectiveness of health and safety aspects of the Queensland Factories and Shops Act, 1896–1931.* Unpublished Bachelors of Administration (Honours) thesis, Griffith University, Brisbane, Australia.

Macrory, R. (2006). *Regulatory justice: Making sanctions effective.* Cabinet Office: Better Regulation Executive, UK.

Magnusson, A. (2012). *Mail from Annica Magnusson.* OHS officer Vårdförbundet, February 27, 2012.

Malenfant, R. (2009). Risk, control and gender: Reconciling production and reproduction in the risk society. *Organization Studies, 30*(2/3): 205–226.

Malenfant, R., LaRue, A., and Vezina, M. (2007). Intermittent work and well-being: One foot in the door, one foot out. *Current Sociology, 55*(6): 814–835.

Mannetje, A. T., Slater, T., McLean, D., Eng, A., Briar, C., and Douwes, J. (2009). *Women's occupational health and safety in New Zealand NOHSAC technical report 13* (pp. 1–166). Wellington, New Zealand: National Occupational Health and Safety Advisory Committee (NOHSAC).

Marmot, M. (1998). Contribution of psychosocial factors to socioeconomic differences in health. *The Millbank Quarterly, 76*(3): 403–448.

Marmot, M., Bosma, H., Hemingway, H., Brunner, E., and Stansfeld, S. (1997). Contribution of job control and other risk factors to social variations in coronary heart disease. *The Lancet, 350*: 235–239.

Marshall, S. (2010). Australian textile clothing and footwear supply chain regulation. In C. Fenwick and T. Novitz (Eds.), *Human rights at work.* Oxford, UK: Hart.

Mathiesen, T. (1981). Disciplining through pulverization. In *Hidden disciplining: Essays in political control.* Oslo, Norway: Author.

Mathiesen, T. (1985). Disziplinierung durch pulverisierung. In T. Mathiesen (Ed.), *Die LautloseDisziplinierung.* AJZ Verlag, Bielefeld, 81.

Mayhew, C., and Quinlan, M. (2006). Economic pressure, multi-tiered subcontracting and occupational health and safety in the Australian long haul trucking industry. *Employee Relations, 28*(3): 212–229.

McBarnet, D. (2007). Corporate social responsibility beyond law, through law, for law: The new corporate accountability. In D. McBarnet, A. Voiculescu, and T. Campbell (Eds.), *The new corporate accountability: Corporate social responsibility and the law.* Cambridge, UK: Cambridge University Press.

McCutchen, B. (2009). *Culture of fear: A report on the status of the enforcement of reprisal protection for workers under the Ontario Occupational Health and Safety Act.* Ontario Federation of Labour.

McDiarmid, M., Oliver, M., Ruser, J., and Gucer, P. (2000). Male and female rate differences in carpal tunnel syndrome injuries: Personal attributes or job tasks? *Environmental Research, 83*(1): 23–32.

McNamara, M., Bohle, P., and Quinlan, M. (2011). Working hours and work life conflict in the hotel industry. *Applied Ergonomics, 42*(2): 225–232.

Menendez, M., Benach, J., Muntaner, C., Amable, M., and O'Campo, P. (2007). Is precarious employment more damaging to women's health than men's? *Social Science & Medicine, 64*: 776–781.

Mergler, D., Brabant, C., Vézina, N., and Messing, K. (1987). The weaker sex? Men in women's working conditions report similar health symptoms. *Journal of Occupational Medicine, 29*: 417–421.

Messing, K. (1998). *One-eyed science: Occupational health and women workers.* Philadelphia, PA: Temple University Press.

Messing, K. (1999). One-eyed science: Scientists, reproductive hazards and the right to work. *International Journal of Health Services, 29*(1): 147–165.

Messing, K. (2002). La place des femmes dans les priorités de recherche en santé au travail au Québec [Women's Place in Workplace Health Research Priorities in Québec]. *Industrial Relations/Relations Industrielles, 57*(4): 660–686.

Messing, K., and Boutin, S. (1997). Les conditions difficiles dans les emplois des femmes et les instances gouvernementales en santé et en sécurité du travail [Difficult conditions in women's jobs and government actors in occupational health and safety]. *Relations Industrielles/ Industrial Relations, 52*(2): 333–362.

Messing, K., Tissot, F., Couture, V., and Bernstein, S. (in press). Conciliation travail-vie personnelle face aux horaires variables et imprévisibles du secteur du commerce de détail au Québec [Work/life balancing in the context of variable and unpredictable schedules in the retail trade sector in Québec]. In D-G. Tremblay (Ed.), *Actes du colloque sur les temps sociaux.* Québec, Canada: Presses de l'Université du Québec.

Messing, K., and de Grosbois, S. (2001). Women workers confront one-eyed science: Building alliances to improve women's occupational health. *Women and Health, 33*(1/2): 125–143.

Messing, K., Dumais, L., Courville, J., Seifert, A. M., and Boucher, M. (1994). Evaluation of exposure data from men and women with the same job title. *Journal of Occupational Medicine, 36*(8): 913–917.

Messing, K., and Elabidi, D. (2003). Desegregation and occupational health: How male and female hospital attendants collaborate on work tasks requiring physical effort. *Policy and Practice in Health and Safety, 1*(1): 83–103.

Messing, K., Seifert, A. M., and Couture, V. (2005). Les femmes dans les métiers nontraditionnels: Le général, le particulier et l'ergonomie. *Travailler, 15*: 131–148.

Michaels, D. (2008). *Doubt is their product.* Oxford, UK: Oxford University.

Milkman, R., Bloom, J., and Harro, V. (Eds.). (2010). *Working for justice: The LA model for organizing and advocacy.* Ithaca, NY: Cornell University Press.

Mills, M. B. (2003). Gender and inequality in the global labor force. *Annual Review of Anthropology, 32*: 41–62.

National Commission on the BP Deepwater Horizon Oil Spill and Offshore Drilling: Report to the President. (2011). *Deep Water: The Gulf oil disaster and the future of offshore drilling.* Available at http://www.oilspillcommission.gov/final-report

National Review into Model Occupational Health and Safety Laws. (2009). *Second Report to the Workplace Relations Ministers' Council.* Canberra: Commonwealth of Australia.

National Standard of Canada. (2013). Psychological Health and Safety in the workplace— Prevention promotion, and guidance to staged implementation, CAN/CSA-Z1003-13/ BNQ 9700-803/2013, BNQ/CSA, Québec, 2013.

Nenonen, S. (2011). Fatal workplace accidents in outsourced operations in the manufacturing industry. *Safety Science, 49*: 1394–1403.

Netterstrom, B., and Hansen, A. (2000). Outsourcing and stress: Physiological effects on bus drivers. *Stress Medicine, 16*: 149–160.

REFERENCES / 237

Netterstrom, B., and Laursen, P. (1981). Incidence and prevalence of ischaemic heart disease among urban bus drivers in Copenhagen. *Scandinavian Journal of Social Medicine*, 9: 75–79.

New York Times. (2010, July 21). Workers on doomed rig voiced concern about safety. Available at http://www.nytimes.com/2010/07/22/us/22transocean.html?pagewanted=1

Nichols, T. (1986, May). Industrial injuries in British manufacturing in the 1980s. *The Sociological Review, 34*(2): 290–306.

Nichols, T. (1997). *The sociology of industrial injury.* London, UK: Mansell.

Nichols, T., and Armstrong, P. (1973). *Safety or profit: Industrial accidents and the conventional wisdom.* Bristol, UK: Falling Wall Press.

Nichols, T., and Armstrong, P. (1976). *Workers divided.* Glasgow, Scotland: Fontana/Collins.

Nichols, T., and Armstrong, P. (1997). Safety or profit? Robens and the conventional wisdom. In T. Nichols (Ed.), *The sociology of industrial injury.* London, UK: Mansell.

Nichols, T. and Beynon, H. (1977). *Living with capitalism.* London, UK: Routledge Kegan Paul.

Nichols, T., and Tucker, E. (2000). OHS management systems in the United Kingdom and Ontario, Canada: A political economy perspective. In K. Frick, P. L. Jensen, M. Quinlan, and T. Wilthagen (Eds.), *Systematic occupational health and safety management.* Oxford, UK: Pergamon Press.

Nichols, T., and Walters, T. (2009). Worker representation on health and safety in the UK—Problems with the preferred model and beyond. In D. Walters and T. Nichols (Eds.), *Workplace health and safety: International perspectives on worker representation.* Basingstoke, UK: Palgrave Macmillan.

Nichols, T., Walters, D., and Tasiran, A. (2008). Trade unions, institutional mediation and industrial safety: Evidence from the UK. *Journal of Industrial Relations, 49*(2): 211–225.

Niedhammer, I., Chastang, J., David, S., Barouhiel, L., and Barrington, G. (2006). Job-strain and effort reward imbalance models in a context of major organizational changes. *International Journal of Occupational and Environmental Health, 12*(2): 111–119.

Nielsen, V. L., and Parker, C. (2009). Testing responsive regulation in regulatory enforcement. *Regulation and Governance, 3*: 376.

NUTEK. (2006). *Näringslivets administrativa kostnader på arbetsrättsområdet.* Stockholm, Sweden: NUTEK.

Nyberg, A., Alfredsson, L., Theorell, T., Westerlund, H., Vahtera, J., and Kivimaki, K. (2009). Managerial leadership and ischaemic heart disease among employees: The Swedish WOLF study. *Occupational and Environmental Medicine, 66*: 51–55.

OHS Insider. (2011a). *BC court: Union's private criminal negligence case can go forward.* Retrieved August 8, 2011 from http://ohsinsider.com/search-by-index/c45/bc-court-union%E2%*0%pps-private-criminal-negligence-prosecution-can-go-forward

OHS Insider. (2011b). Alert: Crown dismisses union brought C-45 case. *OHS Insider* (August 25, 2011). Retrieved September 5, 2011 from http://ohsinsider.com/search-by-index/c45/aug-25-crown-dismisses-union-brought-c-45-case

Olkinuora, M. (1984). Psychogenic epidemics and work. *Scandinavian Journal of Work Environment & Health, 19*: 501–504.

O'Grady, J. (2000). Joint health and safety committees: Finding a balance. In T. Sullivan (Ed.), *Injury and the new world of work.* Vancouver, Canada: UBC Press.

O'Neill, R., Pickvance, S., and Watterson, A. (2007). Burying the evidence: How Great Britain is prolonging the occupational cancer epidemic. *International Journal of Occupational and Environmental Health, 13*(4): 428–436.

238 / SAFETY OR PROFIT?

Ontario Ministry of Labour. (2011). *Safe at work Ontario.* Retrieved April 13, 2011 from http://www.labour.gov.on.ca/english/hs/sawo/index.php

Ontario Ministry of Labour. (2011). *Enforcement statistics 2011.* Available at http://www.labour.gov.on.ca/english/hs/pubs/enforcement/

Osler, D. (2002). *Labour Party PLC. New Labour as a party of business.* Edinburgh, Scotland: Mainstream.

Palan, R. (1998). Trying to have your cake and eating it: How and why the state system has created offshore. *International Studies Quarterly, 42*: 625–644.

Parent-Thirion, A., Fernández Macías, E., Hurley, J., and Vermeylen, G. (2007). *Fourth European Working Conditions Survey.* Luxembourg: European Foundation for the Improvement of Living and Working Conditions.

Parker, C. (2002). *The open corporation: Effective self-regulation and democracy.* Cambridge, UK: Cambridge University Press.

Parker, C. (2004). Restorative justice in business regulation? The Australian Competition and Consumer Commission's use of enforceable undertakings. *Modern Law Review, 67*(2): 209–246.

Parkin, D. (2011). The fraction of cancer attributable to lifestyle and environmental factors in the UK in 2010. *British Journal of Cancer, 105*: 77–81.

Pickvance, S., and O'Neill, R. (2007, July–September). OHS SOS. *Hazards, 99.* Retrieved October 24, 2011 from http://www.hazards.org/workandhealth/ohs.htm

Pikhart, H, Bobak, M., Peasey, A., Pajak, A., Kubinova, R., Malyutina, S., and Marmot, M. (2010). Effort-reward imbalance and all-cause mortality in three post-communist countries: The HAPIEE study. *International Journal of Behavioural Medicine, 17*(Suppl. 1): 136.

Polanyi, K. (2001). *The great transformation: The political and economic origins of our time.* Boston, MA: Beacon Press (first published 1944).

Popp, W., Bruening, T., and Straif, K. (2002). Benufliche Krebserrankungen—Situation in Deutschland. In J. Konietzko and H. Dupuis (Eds.), *Handbuch der Arbeitsmedizin.* Ch. IV. Lundsberg, Sweden: ECOMED Verlag.

Porter, M., and Kramer, M. (2006). Strategy & society. The link between competitive advantage and corporate social responsibility. *Harvard Business Review*, December, 78–92.

Premji, S., Messing, K., and Lippel, K. (2008). "We work by the second!" Piecework remuneration and occupational health and safety from an ethnicity and gender-sensitive perspective. *Pistes, 10*(1): 1–22.

Priest, T. (2010). *Remarks before the National Commission on the BP Deepwater Horizon Oil spill and offshore drilling.* Meeting #2, Panel #3: Regulatory Oversight of Offshore Drilling, August 25. Available at http://www.oilspillcommission.gov/sites/default/files/documents/Tyler%20Priest%20Written%20Statement.pdf

Prieto, N. I. (1997). *Beautiful flowers of the maquiladora: Life histories of women workers in Tijuana.* Austin, TX: University of Texas Press.

Prior, P. F. (1985). Enforcement: An inspectorates view. In W. B. Creighton and N. Gunningham (Eds.), *The industrial relations of occupational health and safety.* Sydney, Australia: Croom Helm.

Probst, I. (2009). La dimension de genre dans la reconnaissance des TMS comme maladies professionnelles. *Revue Pistes, 11*(2). Retrieved February 20, 2012 from http://www.pistes.uqam.v11n2/articles/v11n2a5.htm

Productivity Commission. (2010). *Performance benchmarking of Australian business regulation: Occupational health & safety, research report, 2010.* Canberra, Australia: Productivity Commission.

REFERENCES / 239

Purse, K., and Dorrian, J. (2011). Deterrence and enforcement of occupational health and safety law. *International Journal of Comparative Labour Law and Industrial Relations, 27*: 23–39.

Quandt, S., Grzywacz, J., Marin, A., Carrillo, L., Coates, M., Burke, B., and Arcury, T. (2006). Illnesses and injuries reported by Latino poultry workers in western North Carolina. *American Journal of Industrial Medicine, 49*: 343–351.

Quinlan, M. (1999). The implications of labour market restructuring in industrialized societies for occupational health and safety. *Economic and Industrial Democracy, 20*: 427–460.

Quinlan, M. (2004). Flexible work and organizational arrangements—Regulatory problems and responses. In L. Bluff, N. Gunningham, and R. Johnstone (Eds.), *OHS regulation in the 21st century*. Sydney, Australia: Federation Press.

Quinlan, M. (2007). Organizational restructuring/downsizing, OHS regulation and worker health and well-being. *International Journal of Law and Psychiatry, 30*(4/5): 385–399.

Quinlan, M. (2011). We've been down this road before: Vulnerable work and occupational health in historical perspective. In M. Sargeant and M. Giovanne (Eds.), *Vulnerable workers*. London, UK: Gower.

Quinlan, M. (under review). The pre-invention of precarious employment, ill-health and lessons from history: The case of dockworkers 1880–1945. *International Journal of Health Services*.

Quinlan, M., and Bohle, P. (2004). Contingent work and occupational safety. In J. Barling and M. R. Frone (Eds.), *The psychology of workplace safety*. Washington, DC: American Psychological Association Books.

Quinlan, M., and Bohle, P. (2008). Under pressure, out of control or home alone? Reviewing research and policy debates on the OHS effects of outsourcing and home-based work. *International Journal of Health Services, 38*(3): 489–525.

Quinlan, M., and Bohle, P. (2009). Over-stretched and under reciprocated commitment: Reviewing research on the OHS effects of downsizing and job insecurity. *International Journal of Health Services, 39*(1): 1–44.

Quinlan, M., and Johnstone, R. (2009). The implications of de-collectivist industrial relations laws and associated developments for worker health and safety in Australia, 1996–2007. *Industrial Relations Journal, 40*(5): 426–443.

Quinlan, M., Johnstone, R., and Mayhew, C. (2006). Trucking tragedies: The hidden disaster of mass death in the long haul road transport industry. In E. Tucker (Ed.), *Working disasters*. New York, NY: Baywood.

Quinlan, M., Johnstone, R., and McNamara, M. (2009). Australian health and safety inspectors' perceptions and actions in relation to changed work arrangements. *Journal of Industrial Relations, 51*(4): 559–575.

Quinlan, M., and Mayhew, C. (2000). Precarious employment and workers' compensation. *International Journal of Law and Psychiatry, 22*(5/6): 491–520.

Quinlan, M., Mayhew, C., and Bohle, P. (2001). The global expansion of precarious employment, work disorganization, and consequences for occupational health: A review of recent research. *International Journal of Health Services, 31*(2): 335–414.

Quinlan, M., and Sheldon, P. (2011). The enforcement of minimum labour standards in an era of neoliberal globalisation: An overview. *Economic and Labour Relations Review, 22*(2): 5–32.

240 / SAFETY OR PROFIT?

Quinlan, M., and Sokas, R. (2009). Community campaigns, supply chains and protecting the health and well-being of workers: Examples from Australia and the USA. *American Journal of Public Health, 99*(Suppl. 3): s538–s546.

Quintner, J. L. (1995). The Australian RSI debate: Stereotyping and medicine. *Disability and Rehabilitation, 17*(5): 256–262.

Quist, M. (2011). *Mails and sales statistics on training material from Mats Quist*, market manager at Prevent, October 17–18, 2011.

Rasmussen, J. (2010). *Safety in the making: Studies on the discursive construction of risk and safety in the chemical industry*. Örebro, Sweden: Örebro University.

Reid, J., Ewan, C., and Lowy, E. (1991). Pilgrimage of pain: The illness experiences of women with repetition strain injury and the search for credibility. *Social Science & Medicine, 32*(5): 601–612.

Report of the Royal Commission on the Poor Laws and Relief of Distress. (1909). Presented to both Houses of Parliament, British House of Commons Parliamentary Papers. [Cmnd 4499]. London, UK: HMSO.

Rhodes, G. (1981). *Inspectorates in British Government: Law enforcement and standards of efficiency*. London, UK: Allen and Unwin.

Robens, A. (1972). *Safety and health at work: Report of the committee 1970–1972*, Cmnd 5034, London, UK: HMSO.

Rogers, D. W. (2009). *Making capitalism safe*. Urbana and Chicago: University of Illinois Press

Rothstein, B. (1992). *Den korporativa staten*. Stockholm, Sweden: Norstedts.

Roy, M. (2003). Self-directed work teams and safety: A winning combination? *Safety Science, 41*: 359–376.

Royal Commission on Labour. (1894). *Fifth and final report of the Royal Commission on Labour. Part I* (C 7421), London, UK: HMSO, British House of Commons Parliamentary Papers.

Rushton, L., Bagga, S., Bevan, R., Brown, R., Cherrie, J., Holmes, P., et al. (2010). *The burden of occupational cancer in Great Britain overview report*. Norwich, UK: HSE Books.

SACO. (2010). *Färdriktning 2010–2013*. Stockholm, Sweden: SACO.

Safety and Health Practitioner (SHP). (2009). *Shift work and cancer*. April 7.

Sargeant, M., and Tucker, E. (2009). Layers of vulnerability in occupational safety and health for migrant workers: Case studies from Canada and the UK. *Policy and Practice in Health and Safety, 7*(2): 51–74.

Schell, G. (2002) Farmworker exceptionalism under the Law. In C. D. Thompson and M. F. Wiggins (Eds.), *Farmworkers' lives, labor and advocacy*. Austin, TX: University of Texas Press.

Schofield, T., Reeve, B., and McCallum, R. (2009). Deterrence and OHS prosecutions: Prosecuted employers' responses. *Journal of Occupational Health and Safety—Australia and New Zealand, 25*: 263–276.

Schweder, P. (2009). *OHS experiences of seasonal direct-hire temporary employees*. PhD thesis, University of New South Wales.

Seifert, A., and Messing, K. (2006). Cleaning up after globalization: An ergonomic analysis of the work activity of hotel cleaners. *Antipode, 38*(3): 557–578.

Seifert, A. M., Messing, K., Riel, J., and Chatigny, C. (2007). Precarious employment conditions affect work content in education and social work: Results of work analyses. *International Journal of Law and Psychiatry, 30*(4/5): 299–310.

REFERENCES / 241

Seixas, N. S., Blecker, H., Camp, J., and Neitzel, R. (2008). Occupational health and safety experience of day laborers in Seattle, WA. *American Journal of Industrial Medicine,* 51(6): 399–406.

Senior Labour Inspectors' Committee (SLIC). (2012). *EU strategic priorities 2013–2020.* A submission from SLIC. European Commission, DGV, Luxembourg.

Shirom, A., Toker, S., Berliner, S., and Shapira, I. (2008). The job demand-control-support model and stress related low-grade inflammatory responses amongst healthy employees: A longitudinal study. *Work and Stress,* 22(2): 138–152.

Silbey, S. S. (2009). Taming Prometheus: Talk about safety culture. *Annual Review of Sociology,* 35: 341–369.

Siegrist, J. (1996). Adverse health effects of high-effort/low-reward conditions. *Journal of Occupational Health Psychology,* 1(1): 27–41.

Sigler, J. A., and Murphy, J. E. (1988). *Interactive corporate compliance.* New York, NY: Quorum Books.

Sigler, J. A., and Murphy, J. E. (1991). *Corporate law breaking and interactive compliance: Resolving the regulation-deregulation dichotomy.* New York, NY: Quorum Books.

Silverstein, M. (2007). Ergonomics and regulatory politics: The Washington State case. *American Journal of Industrial Medicine,* 50: 391–401.

Silvestre, J. (2010). Improving workplace safety in the Ontario manufacturing industry, 1914–1939. *Business History Review,* 84: 527–550.

Simpson, R. (2000). Presenteeism and the impact of long hours on managers. In D. Winstanley and J. Woodall (Eds.), *Ethical issues in contemporary human resource management.* Basingstoke, UK: Macmillan Business.

Sjöholm, B. (2010). Telephone interview with Börje Sjöholm, OHS officer, Unionen, December 6, 2010.

Sjöström, J. (2006). *Worker participation in occupational health and safety management.* Paper to the conference on Working with Machines, Vadstena, September 4–6. Linköping, Sweden: Department of Technology and Social Change, Linköping University.

Skagerfält, M. (2011). Telephone interview with Magnus Skagerfält, OHS officer Sveriges Ingenjörer October 3, 2011 and mail September 9, 2011.

Slapper, G. (2000). *Blood in the bank. Social and legal aspects of death at work.* Aldershot, UK: Ashgate.

Slapper, G., and Tombs, S. (1999). *Corporate crime.* London, UK: Longman.

Smith, C. K., Silverstein, B. A., Bonauto, D. K., Adams, D., and Fan, Z. J. (2010). Temporary workers in Washington State. *American Journal of Industrial Medicine,* 53(2): 135–145.

Smith, D. (2000). *Consulted to death.* Winnipeg, Canada: Arbeiter Ring.

Smith, P. R. (2011). The pitfalls of home: Protecting the health and safety of paid domestic workers. *Canadian Journal of Women and the Law,* 23(1): 309–339.

SOU. (1972). *Bättre Arbetsmiljö—Delbetänkande Avgivet Av Arbetsmiljöutredningen.* Stockholm, Sweden: Arbetsmarknadsdepartementet.

SOU. (2007). *Bättre Arbetsmiljöregler II—Skyddsombud, Beställaransvar, Byggarbetsplatser.* Stockholm, Sweden: Fritzes.

SOU. (2009). *Marknadsorienterade Styrmedel på Arbetsmiljöområdet.* Stockholm, Sweden: Fritzes.

Standing, G. (2011). *The Pecariat: The new dangerous class.* London, UK: Bloomsbury.

242 / SAFETY OR PROFIT?

Statoil. (2010). *Gullfaks incident entails drill routine changes.* Retrieved November 5 from http://www.statoil.com/en/NewsAndMedia/News/2010/Pages/05NovGullfaksC_report.aspx

Stegenga, J., Bell, E., and Matlow, A. (2002). The role of nurse understaffing in nosocomial viral gastrointestinal infections on a general pediatrics ward. *Infection Control and Hospital Epidemiology, 23*(3): 133–139.

Steinberg, M. (2004). *Skyddsombud—I allas Intresse.* Stockholm, Sweden: Norstedts Juridik.

Steinzor, R. (2011). Lessons from the North Sea: Should "safety cases" come to America? *Boston College Environmental Affairs Law Review, 38*: 417–444. Available at http://papers.ssrn.com/sol3/papers.cfm?abstract_id=1735537

Steinzor, R., and Havemann, A. (2011). Too big to obey: Why BP should be debarred. *William & Mary Environmental Law and Policy Review, 36*: 81–118.

Stellman, J. M. (1978). *Women's work, women's health: Myths and realities.* New York, NY: Pantheon.

Stinson, J., Pollack, N., and Cohen, M. (2005). *The pains of privatization.* British Columbia, Canada: Canadian Centre for Policy Alternatives.

Stirling University. (2008). *International Conference on Occupational and Environmental Carcinogen.* Papers at http://www.nmh.stir.ac.uk/research/occupational-environmental-cancer.php

Stock, S., Funes, A., Délisle, A., St-Vincent, M., Turcot, A., and Messing, K. (2011). Troubles musculo squelettiques. In M. Vézina, E. Cloutier, S. Stock, K. Lippel, É. Fortin, A. Delisle, M. St-Vincent, A. Funes, P. Duguay, S. Vézina, and P. Prud'homme (Eds.), *Enquête Québécoise sur des conditions de Travail, d'Emploi et de Santé et de Sécurité du Travail (EQCOTESST)* (pp. 499–530). Montréal, Canada: Institut de recherche Robert Sauvé en santé et sécurité du travail, Institut national de santé publique du Québec, Institut de la Statistique du Québec.

Stone, K. V. W. (2006). Flexibilization, globalization, and privatization: Three challenges to labour rights in our time. *Osgoode Hall Law Journal, 44*: 77–104.

Storey, R. (2005). Activism and the making of occupational health and safety law in Ontario, 1960s–1980. *Policy and Practice in Health and Safety, 3*(1): 41–68.

Storey, R., and Tucker, E. (2006). All that is solid melts into air: Worker participation and occupational health and safety regulation in Ontario, 1970–2000. In V. Mogensen (Ed.), *Worker safety under siege.* Armonk, NY: M. E. Sharpe.

Straif, K. (2007). Carcinogenicity of shift-work, painting, and fire-fighting. *The Lancet Oncology, 8*(12): 1065–1066.

Straif, K. (2008). The burden of occupational cancer. *Occupational and Environmental Medicine, 65*: 787–788.

Sundberg, M. (2011). Mail from Mats Sundberg, training officer at LO, October 19, 2011.

Sveriges Läkarförbund. (2005). *Vägledning för Skyddsombud. Råd om hur Läkares Arbetsmiljö kan Förbättras.* Stockholm, Sweden: Sveriges Läkarförbund.

Swerdlow, A. (2003). Shift work and breast cancer: A critical review of the epidemiological evidence. Research Report 132. Norwich, UK: HSE Books.

Taiwo, O. A., Cantley, L. F., Slade, M. D., Pollack, K. M., Vegso, S., Fiellin, M. G., and Cullen, M. R. (2009). Sex differences in injury patterns among workers in heavy manufacturing. *American Journal of Epidemiology, 169*(2): 161–166.

Takala, J. (2008). *Cancer prevention and EASHW.* Stirling University International Conference on Occupational and Environmental Carcinogen. Papers at http://www.nmh.stir.ac.uk/research/occupational-environmental-cancer.php

Taylor, H. (2001). Insights into participation from critical management and labour process perspectives. In B. Cooke and U. Kothari (Eds.), *Participation: The new tyranny?* London, UK: Zed Books.

TCO. (2011). *Jobben och Jobbet—Om Jobbens Betydelse och Arbetets Villkor*. Stockholm, Sweden: TCO.

Teubner, G. (1983). Substantive and reflexive elements in modern Law. *Law and Society Review, 17*: 239–285.

Tombs, S. (1999). Death and work in Britain. *Sociological Review, 47*(2): 345–367.

Tombs, S., and Whyte, D. (2006). Work and risk. In S. Walklate and G. Mythen (Eds.), *Beyond the risk society: Critical reflections on risk and human security*. Buckingham, UK: Open University Press.

Tombs, S., and Whyte, D. (2007). *Safety crimes*. Devon, UK: Willan Publishing.

Tombs, S., and Whyte, D. (2008). *The crisis in enforcement: The decriminalisation of death and injury at work*. London, UK: Crime and Society Foundation.

Tombs, S., and Whyte, D. (2010a). A deadly consensus: Worker safety and regulatory degradation under New Labour. *British Journal of Criminology, 50*(1): 46–65.

Tombs, S., and Whyte, D. (2010b). *Regulatory surrender: Death, injury and the non-enforcement of law*. London, UK: Institute of Employment Rights.

Tompa, E., Trevithick, S., and McLeod, C. (2007). Systematic review of the prevention incentives of insurance and regulatory mechanisms for occupational health and safety. *Scandinavian Journal of Work and Environmental Health, 33*(2): 85–95.

Townsend, P. (1979). *Poverty in the UK*. Harmondsworth, UK: Penguin.

Toxics Use Reduction Institute (TURI). (2012). Success stories: TURA 20th anniversary leaders reduce toxic chemical use by 3 million pounds. *University of Massachusetts, Lowell*. Retrieved February 23, 2012 from http://www.turi.org/About/Success_Stories

Trade Unions Congress (TUC). (2012). *Occupational cancers—The figures: Briefing for activists—February 2012*. Retrieved February 24, 2012 from http://www.tuc.org.uk/workplace/tuc-20569-f0.cfm

Tucker, E. (1990). *Administering danger in the workplace: The law and politics of occupational health and safety legislation in Ontario, 1850–1914*. Toronto, Canada: University of Toronto Press.

Tucker, E. (1992). Worker participation in health and safety regulation: Lessons from Sweden. *Studies in Political Economy, 37*: 95–127.

Tucker, E. (2003). Diverging trends in worker health and safety protection and participation in Canada, 1985–2000. *Relations Industrielles/Industrial Relations, 58*: 395–426.

Tucker, E. (Ed.). (2006a). *Working disasters: The politics of recognition and response*. Amityville, NY: Baywood.

Tucker, E. (2006b). Will the vicious circle of precariousness be unbroken?: The exclusion of Ontario farm workers from the Occupational Health and Safety Act. In L. Vosko (Ed.), *Precarious employment: Understanding labour market insecurity in Canada*. Montreal and Kingston, Canada: McGill-Queen's University Press.

Tucker, E. (2007). Re-mapping worker citizenship in contemporary occupational health and safety regimes. *International Journal of Health Services, 37*: 145–170.

Tucker, E. (2010). Renorming labour law: Can we escape labour law's recurring regulatory dilemmas? *Industrial Law Journal, 39*: 99–138.

Tuohy, C., and Simard, M. (1993). *The impact of JHSC in Ontario and Quebec*. A study prepared for the Canadian Association of Administrators of Labour Law.

244 / SAFETY OR PROFIT?

Underhill, E. (2008). *Double jeopardy: Occupational injury and rehabilitation of temporary agency workers*. Unpublished PhD thesis, University of New South Wales.

Underhill, E., and Quinlan, M. (2011). How precarious employment affects health and safety at work: The case of temporary agency workers. *Relations Industrielles/Industrial Relations, 66*(3): 397–421.

Unionen. (2008). *Arbetsmiljöbarometern, November 2008*. Stockholm, Sweden: Unionen.

Unionen. (2010a). *Arbetsmiljöbarometern 2010 del 2—Arbetsplatser utan arbetsmiljöombud*. Stockholm, Sweden: Unionen.

Unionen. (2010b). *Arbetstid är mer än pengar. Arbetsmiljöbarometern November 2010*. Stockholm, Sweden: Unionen.

Upmark, M., Borg, K., and Alexanderson, K. (2007). Gender differences in experiencing negative encounters with healthcare: A study of long-term sickness absentees. *Scandinavian Journal of Public Health, 35*: 577–584.

Uppal, S. (2011). *Unionization 2011*. Ottawa: Statistics Canada. Retrieved February 23, 2012 from http://www.statcan.gc.ca/pub/75-001-x/2011004/article/11579-eng.pdf

Upstream. (2012). *Total in North Sea emergency*, Retrieved March 26 from http://www.upstreamonline.com/live/article1242808.ece

U.S. Department of the Interior. (2010). *Salazar announces regulations to strengthen drilling safety, reduce risk of human error on offshore oil and gas operations*. Retrieved September 30 from http://www.doi.gov/news/pressreleases/Salazar-Announces-Regulations-to-Strengthen-Drilling-Safety-Redu.cfm

Vahtera, J., Kivimaki, M., Forma, P., Wikstrom, J., Halmeenaki, T., Linna, A., and Pentti, J. (2005). Organizational downsizing as a predictor of disability pension: The 10-town prospective cohort study. *Journal of Epidemiology and Community Health, 59*: 238–242.

Vahtera, J., Kivimaki, M., and Pentti, J. (1998). Effects of organizational downsizing on health of employees. *The Lancet, 350*: 1124–1128.

van den Abeele, E. (2009). The Better Regulation agenda: A new deal in the building of Europe? ETUI Policy Brief. *European Social Policy, 1*. Brussels, Belgium: ETUI.

van den Abeele, E. (2010). *The European Union's better regulation agenda*, Report 112. Brussels, Belgium: European Trade Union Institute.

Verheugen, G. (2008, June 20). *Reducing red tape for Europe*. Brussels, Belgium.

Vézina, M., Cloutier, E., Stock, S., Lippel, K., Fortin, É., Delisle, A., and Prud'homme, P. (2011). *Québec survey on working and employment conditions and occupational health and Safety (EQCOTESST)*. Montreal, Canada: Institut de recherche Robert Sauvé en santé et sécurité du travail, Institut national de santé publique du Québec, Institut de la Statistique du Québec.

Vézina, M., Stock, S., Funes, A., Delisle, A., St-Vincent, M., Turcot, A., and Arcand, R. (2011). Description de l'environnement organizationnel et des contraintes physiques à l'emploi principal. In M. Vézina, E. Cloutier, S. Stock, K. Lippel, É. Fortin, A. Delisle, M. St-Vincent, A. Funès, P. Duguay, S. Vézina, and P. Prud'homme (Eds.), *Enquête Québécoise sur des Conditions de Travail, d'Emploi et de Santé et de Sécurité du Travail (EQCOTESST)* (pp. 223–332). Montréal, Canada: Institut de recherche Robert Sauvé en santé et sécurité du travail, Institut national de santé publique du Québec, Institut de la Statistique du Québec.

Vézina, N., Tierney, D., and Messing, K. (1992). When is light work heavy? Components of the physical workload of sewing machine operators which may lead to health problems. *Applied Ergonomics, 23*: 268–276.

REFERENCES / 245

Virtanen, M., Kivimaki, M., Joensuu, M., Virtanen, P., Elovainio, M., and Vahtera, J. (2005). Temporary employment and health: A review. *International Journal of Epidemiology, 34*: 610–622.

Vives, A., Vanroellen, C., Amable, M., Ferrer, M., Moncada, S., Llornes, C., Muntaner, C., Benavides, F., and Benach, J. (2011). Employment precariousness in Spain: Prevalence, social distribution, and population-attributable risk percent of poor mental health. *International Journal of Health Services, 41*(4): 625–646.

Vogel, L. (2009). Special report: Better regulation. *HESA. Newsletter of the Health and Safety Department of the European Trade Union Institute, 35*: 3–15, Brussels, Belgium: ETUI.

Vogel, L., and van den Abeele, E. (2010). *Better regulation: A critical assessment,* Report 113. Brussels, Belgium: ETUI.

Vogel, L., and Walters, D. (2009). An afterword on European Union policy and practice. In D. Walters and T. Nichols (Eds.), *Workplace health and safety: International perspectives on worker representation.* Basingstoke, UK: Palgrave Macmillan.

Vosko, L. F. (2010). *Managing the margins—Gender, citizenship, and the international regulation of precarious employment.* New York, NY: Oxford University Press.

Wallen, J., Waitzkin, H., and Stoeckle, J. (1979). Physician stereotypes about female health and illness. *Women & Health, 4*(2): 135–146.

Walters, D. (2001). *Health and safety in small enterprises: European strategies for managing improvement.* Brussels, Belgium: PIE-Peter Lang.

Walters, D. (Ed.). (2002). *Regulating health and safety management in the European Union.* Brussels, Belgium: Peter Lang.

Walters, D. (2003). *Workplace arrangements for OHS in the 21st century.* National Research Centre for OHS Regulation, Working Paper 10. The Australian National University.

Walters, D., and James, P. (2009). *Understanding the role of supply chains in influencing health and safety at work.* Report prepared for Institute of Occupational Safety and Health. Leicester, UK: IOSH.

Walters, D., and James, P. (2011). What motivates employers to establish preventive management arrangements within supply chains? *Safety Science, 49*: 988–994.

Walters, D., Johnstone, R., Frick, K., Quinlan, M., Gringras, G., and Thebaud-Mony, A. (2011). *Regulating work risks: A comparative study of inspection regimes in times of change.* Cheltenham, UK: Edward Elgar.

Walters, D., and Nichols, T. (2007). *Worker representation and workplace health and safety.* Basingstoke, UK: Palgrave Macmillan.

Walters, D., and Nichols, T. (Eds.). (2009). *Workplace health and safety: International perspectives on worker representation.* Basingstoke, UK: Palgrave Macmillan.

Watterson, A. (1993). Occupational health in the UK gas industry. In S. Platt, H. Thomas, S. Scott, and G. Williams (Eds.), *Locating health.* London, UK: Avebury Press.

Watterson, A. (1999a, January–March). Research for truth. *Hazards, 65*: 14–15.

Watterson, A. (1999b). Why we still have old epidemics and endemics in occupational health: Policy and practice failures and some solutions. In N. Daykin and L. Doyal (Eds.), *Health and work: Critical perspectives.* London, UK: Macmillan.

Watterson, A. (2009). While you were sleeping. *Hazards, 106*: April–June. Retrieved October 24, 2011 from http://www.hazards.org/hours/shiftwork.htm

Watterson, A. (2012). Work and wider environmental cancer prevention: UK perspectives. *Scottish Cancer Prevention Newsletter, 3*(1): 6. Dundee University, Scotland.

Watterson, A., Gorman, T., Malcolm, C., Robinson, M., and Beck, M. (2006). The economic costs of health service treatments for asbestos-related mesothelioma deaths. *Annals of New York Academy Science, 1076*: 871–881.

Watterson, A., Gorman, T., and O'Neill, R. (2008). Occupational cancer prevention in Scotland: A missing public health priority. *European Journal of Oncology, 13*(3): 161–169.

Weil, D. (2010). *Improving workplace conditions through strategic enforcement.* Report to the Wage and Hour Division, May 2010. Retrieved August 8, 2011 from http://papers.ssrn.com/sol3/papers.cfm?abstract_id=1623390

Weil, D., and Pyles, A. (2005). Why complain? Complaints, compliance, and the problem of enforcement in the U.S. workplace. *Comparative Labor Law and Policy Journal, 27*: 59–92.

Wenger, E. (1998). *Communities of practice. learning, meaning, and identity.* Cambridge, UK: Cambridge University Press.

WHO. (2006). *Prevention of occupational cancer.* World Health Organization. GOHNET newsletter, No 11. Available at http://www.who.int/occupational_health/publications/newsletter/gohnet11e.pdf

WHO. (2009). *Global health risks: Mortality and burden of disease attributable to selected major risks.* Geneva, Switzerland: World Health Organization.

WHO. (2011). *Interventions for primary prevention. Asturias Declaration: A call to action.* International Conference on Environmental and Occupational Determinants of Cancer. Retrieved June 9, 2011 from http://www.who.int/phe/news/events/inter?national_conference/Call_for_action_en.p?df

Wikman, A. (2010). Changes in power, influence and organization in Sweden. In S. Marklund and A. Härenstam (Eds.), *The dynamics of organizations and healthy work.* Arbetsliv i omvandling 2010: 5. Växjö, Sweden: Linnéuniversitet.

Williams, A. (1969). *Life in a railway factory.* New York, NY: Augustus M Kelley. (Original work published 1915)

Williamson, A. (2007). Predictors of psychostimulant use by long distance truck drivers. *American Journal of Epidemiology, 166*(11): 1320–1326.

Wilson, C. M., Douglas, K. S., and Lyon, D. R. (2011). Violence against teachers: Prevalence and consequences. *Journal of Interpersonal Violence, 26*(12) 2353: 1–19.

Wise, J. (2009). Danish night shift workers awarded compensation. *British Medical Journal, 338*: 1152.

Woolfson, C., and Beck, M. (2003). Occupational health and safety in transitional Lithuania. *Industrial Relations Journal, 34*: 241–259.

Woolfson, C., and Beck, M. (2004). The Employment Relations Act 1999 and partnership union recognition in Britain's offshore oil and gas industry. *Industrial Relations Journal, 35*(4): 344–358.

Woolfson, C., and Beck, M. (2005). Corporate social responsibility in the international oil industry. In C. Woolfson and M. Beck (Eds.), *Corporate responsibility failures in the oil industry.* Amityville, NY: Baywood.

Woolfson, C., Foster, J., and Beck, M. (1996). *Paying for the Piper: Capital and labour in Britain's offshore oil industry.* London, UK: Routledge/Curzon.

Workplace Relations Ministers' Council. (2008). *Performance monitoring reports, comparison of occupational health and safety and workers' compensation in Australia and New Zealand* (10th ed.). Canberra, Australia: Safe Work Australia.

Workplace Relations Ministers' Council. (2010). *Performance monitoring reports, comparison of occupational health and safety and workers' compensation in Australia and New Zealand* (12th ed.). Canberra, Australia: Safe Work Australia.

Wright, E. O. (2000). Working-class power, capitalist-class interests, and class compromise. *American Journal of Sociology, 105*: 957–1002.

Wright, M., Marsden, S., and Antonelli, A. (2004). *Building an evidence base for the Health and Safety Commission strategy to 2010 and beyond: A literature review of interventions to improve health and safety compliance.* Norwich, UK: HSE Books.

Young, D. (2010). *Common sense, common safety. A report by Lord Young of Graffham to the Prime Minister following a Whitehall-Wide review of the operation of health and safety laws and the growth of the compensation culture.* London, UK: The Cabinet Office. Available at http://www.number10.gov.uk/wp-content/uploads/402906_CommonSense_acc.pdf

Zahm, S. H., and Blair, A. (2003). Occupational cancer among women: Where have we been and where are we going? *American Journal of Industrial Medicine, 44*(6): 565–575.

Zahm, S. H., Pottern, L. M., Lewis, D. R., Ward, M. H., and White, D. W. (1994). Inclusion of women and minorities in occupational cancer epidemiologic research. *Journal of Occupational Medicine, 36*(8): 842–847.

Zoller, H. M. (2003). Health on the line: Identity and disciplinary control in employee occupational health and safety discourse. *Journal of Applied Communication Research, 31*: 118–139.

Contributors

KAJ FRICK is professor of work science, with a speciality in OHS management. He worked at the Swedish National Institute for Working Life but is currently at Malardalen University, Sweden. Since 1975, he has planned, coordinated, executed, and promoted research on the organization, regulation, and industrial relations of OHS management, often with an international perspective. He has also written on the economic discourse of OHS, comparative OHS politics, the regulation of OHSM, voluntary OHSM systems, worker representation and participation in OHS management, and OHS management in small firms.

RICHARD JOHNSTONE is professor at the Griffith Law School, Australia, where he was the Director of the Centre for Socio-Legal Research from 2003 to early 2010. From August 2001 until April 2004, he was the foundation director of the National Research Centre for Occupational Health and Safety Regulation, based in the Regulatory Institutions Network at the Australian National University. Since 2007, he has been one of the two co-directors of that center. Richard is also currently adjunct professor at the Australian National University and honorary professor in the Department of Ageing, Work and Health at the University of Sydney. Richard's academic interests are in regulation (particularly occupational health and safety regulation), sociolegal research, labor law, and legal education. Richard's books on occupational health and safety regulation include *Occupational Health and Safety Law and Policy* (2nd ed., 2004); *Regulating Workplace Safety: Systems and Sanctions* (with Neil Gunningham) (1999), *Occupational Health and Safety, Courts and Crime: The Legal Construction of Occupational Health and Safety Offences in Victoria,* (2003), and *OHS Regulation for a Changing World of Work* (with E. Bluff and N. Gunningham [eds.]) (Federation Press, 2004).

WAYNE LEWCHUK is professor of labour studies and economics at McMaster University, Hamilton, Ontario. He is currently the graduate chair of the MA in work and society. His early research interests focused on the history of technology in the automobile industry. His more recent work examines work reorganization in a variety of sectors, including auto and the public sector and its impact on the quality of life in work and health outcomes. He was funded by the Ontario Workplace Safety and Insurance Board and the Lupina Foundation to study the impact of precarious employment relationships on health outcomes. The results of this study are forthcoming in a volume titled *Working Without Commitments: Precarious Employment and Health*, to be published by McGill Queen's University Press. He is currently the co-director of a 5-year joint university/community research program on Poverty and Employment Precarity in Southern Ontario (PEPSO). He holds a BA and MA in economics from the University of Toronto and a PhD in economics from the University of Cambridge.

KATHERINE LIPPEL is currently full professor of law at the Faculty of Law (Civil Law Section) at the University of Ottawa and holds the Canada Research Chair in Occupational Health and Safety Law. She is also associate professor of law at the Université du Québec à Montréal, where she was a professor from 1982 to 2006, and a member of the CINBIOSE research center, an adjunct professor in Carleton University's School of Social Work, and an adjunct scientist at the Institute for Work & Health. She specializes in legal issues relating to occupational health and safety and workers compensation, and is the author of numerous articles and books in the field. Her research interests include work and mental health; health effects of compensation systems; policy, precarious employment, and occupational health; interactions between law and medicine in the field of occupational health and safety; disability prevention and compensation systems; women's occupational health; regulatory issues in occupational health and safety; globalization and occupational health and safety.

KAREN MESSING is professor emerita in the Department of Biological Sciences at the University of Quebec at Montreal. Her research deals with applications of gender-sensitive analysis in occupational health, constraints and demands of work in the health care and service sectors, especially prolonged static standing and work/family articulation and regulation. She has authored over 120 peer-reviewed published scientific articles. She wrote *One-Eyed Science: Occupational Health and Working Women* (1998) and, with Piroska Ostlin, the WHO's *Gender Equality, Work and Health: A Review of the Evidence* (2006). She was co-founder and first director of CINBIOSE, an environmental health research center involved in many collaborations with unions and other community organizations. She founded, chaired, and now co-chairs the Gender and Work Technical Committee of the International Ergonomics Association. She has received the Governor General of Canada's Award in Commemoration of the Persons Case (2009) and the Jacques-Rousseau Prize for interdisciplinary research from the Francophone Science Association (1993). She received the Distinguished Fellowship of the Quebec Council for Social Research in 1995–1997 and the Distinguished Fellowship of the Canadian Institutes for Health Research in 2001–2006. She is an honorary professor at the Institute for Feminist Studies and Research at the Université du Québec à Montréal.

THEO NICHOLS was formerly distinguished research professor Cardiff School of Social Sciences, UK where he is now Professor Emeritus and an associate researcher at CWERC. He has written widely on a variety of subjects in the general field of economic sociology, including class relations, management, and productivity, and has a special interest in labor relations in Turkey and China. He was one of the first sociologists in the UK to research health and safety at work, a field to which he has returned at various times since the publication with Peter Armstrong of *Safety or Profit?* in 1973.

MICHAEL QUINLAN has been researching regulatory aspects of employment, especially OHS, for over 30 years. His particular interests in precarious employment, immigrant, foreign, and vulnerable workers and labor history are internationally acclaimed. In 2000–2001, he undertook an inquiry for the Motor Accidents Authority of New South Wales into safety in the long-haul trucking industry. More recently, he was involved in the inquiry into the Beaconsfield mining disaster. His ongoing research includes an investigation of the OHS regulatory implications of precarious employment and research assessing policy and practice in the move to process standards in OHS internationally. He has served on a number of government bodies providing advice on OHS policy and regulatory issues and prepared reports and submissions to numerous government inquiries in regulatory aspects of employment.

STEVE TOMBS has been professor of criminology at The Open University since 2013 and was formerly Professor of Sociology at Liverpool John Mores University where he taught in the Schools of Business, Law, and Social Science. He was previously a lecturer in sociology at the University of Wolverhampton, where he gained his PhD in 1991. He had previously completed his MA (Econ.) in political theory at the University of Manchester. He has a long-standing interest in the incidence, etiology, and regulation of corporate crime and harm, drawing on criminology, sociology, social and public policy, and political economy. He is currently a member of the Editorial Board of State Crime and Policy and Practice in Health and Safety, and on the steering committee of the European Group for the Study of Deviance and Social Control. Steve has enjoyed long-term working relationships with various campaigning organizations, including the Hazards movement, the Institute for Employment Rights and Families Against Corporate Killers, numerous trades unions, and the local collective Catalyst Media. In 1999, he co-founded the Centre for Corporate Accountability, and was chairman of its Board of Directors. He currently edits a column—Crimes of the Powerful and Insurgent Resistance—with Frank Pearce on Crime Talk.

ERIC TUCKER is professor at Osgoode Hall Law School, York University, Toronto. He has published extensively on the history and current state of labor and employment law and occupational health and safety regulation. He is the author of *Administering Danger in the Workplace* (1990) and co-author of *Labour Before the Law* (2001) and *Self-Employed Workers Organize* (2005). He is also the editor of *Working Disasters: The Politics of Recognition and Response* (2006) and co-editor of *Work on Trial: Canadian Labour Law Struggles* (2010).

DAVID WALTERS is professor of work environment and Director of the Cardiff Work Environment Research Centre (CWERC), a Cardiff University research center in the School of Social Sciences. His research and writing is on various aspects of the work environment, and he has particular interests in employee representation and consultation on health and safety, the politics of health and safety at work, regulating health and safety management, chemical risk management at work, and health and safety in small firms. Recent publications include *Regulating Workplace Risks: A Comparative Study of Inspection Regimes in Times of Change* (2011), *Workplace Health and Safety: International Perspectives on Worker Representation* (2009), and *Within Reach? Managing Chemical Risks in Small Enterprises* (2008). He is the editor of the international journal, *Policy and Practice in Health and Safety*, a member of the IOSH Research Committee, and has advised several state inquiries on health and safety.

ANDREW WATTERSON is professor in the School of Health at Stirling University in Scotland, head of its Occupational and Environmental Health Research Group, and director of its Centre for Public Health and Population Health Research. During the 1970s, he worked for the WEA and for the General, Municipal and Boiler Makers Trade Union. During the 1980s, he completed a PhD part-time at Bristol University under the supervision of Theo Nichols. His main research interests currently relate to occupational and environmental health, particularly in the agricultural, electronics, and fish farming industries, using action-research and health impact assessment methods.

DAVID WHYTE is reader in sociology at the University of Liverpool, where he teaches and researches state and corporate power. He has written extensively on safety crimes, death and injury at work, and enforcement issues. His books on the regulation of workplace safety include *Safety Crimes* (2007) and *Regulatory Surrender* (2010). He was also co-author of "ICL/Stockline Disaster: An Independent Report on Working Conditions Prior to the Explosion" (2007) and is a former member of the Scottish Government Expert Group on Corporate Homicide. His study of the role of corporations in the reconstruction and corruption of the Iraqi economy won the 2007 Radinowicz Prize for Criminology.

CHARLES WOOLFSON is currently professor of labour studies at the Institute for Research on Migration, Ethnicity and Society (REMESO) at Linköping University, Sweden. He was previously a full-time Marie Curie Chair in the Baltic States (2004–2007), during which period he researched labor standards in the new accession EU member states.

Index

Apathy, 3, 12, 44, 71, 96, 157, 158, 163, 180

Business process engineering, 18

Carcinogens, 39, 100, 137-156
Child labour, 19, 21, 22, 23
Children, 19, 20
Compensation
compensation culture, 7-8, 110
workers' compensation, 25, 30, 32, 37, 41, 42, 43, 45-48, 54, 73, 74, 89, 90, 137, 148, 149, 150, 151, 153, 154, 203
Consultation (*see* Workers representation and consultation
Corporate manslaughter/homicide, 125, 130, 133, 209
Corporate social responsibility/codes of conduct, 7, 84, 89, 107-109
Countries
Australia, 11, 15, 19, 37, 39, 42, 49, 68, 84, 85, 113-132, 151, 208, 209, 210
Brazil, 23
Canada, 33-48, 71-95, 135, 146, 154, 155, 157-180
Denmark, 53, 147, 149, 154
France, 7, 23, 37, 48, 139, 154
Germany, 7, 138, 147
Italy, 140, 196
Luxemburg, 140
Mexico, 35
Netherlands, 211
Norway, 27, 202
Romania, 148
Spain, 25, 155

[Countries]
Sweden, 9, 49, 51-70, 159
United Kingdom, 3-8, 10-11, 97, 111, 137-156, 184-185, 190-192, 199-200
United States, 34, 37, 39, 41, 52, 211, 212, 215
Criminal liability, 103, 128-130, 131
Cullen Report, 134, 184, 185, 191, 192, 193, 196, 199
"Culture," 6, 8, 174, 180

Dean Report, 90, 93, 160-162, 215
Decasualisation of labour, 18
Decriminalisation of occupational health and safety, 11, 113-133
Deepwater Horizon, 12, 13, 23, 134, 181-216
Defence counsel techniques, 117-121
De-territorialization, 203

Employers
duties, 2, 50-52
and precarious work, 26-27, 29
small, 24, 54, 58-59, 64, 69, 93, 160
underground, 161, 164
and women workers, 34-37, 46, 48
Employment contracts
agency, 6, 24, 27, 29, 37, 57, 61, 69, 160, 169, 171, 176, 206
casual, 18, 38, 54, 59, 62
part time, 28, 38, 171
permanent/temporary, 24, 25, 28, 54, 61, 84
self-employed, 6, 23, 38, 54, 160, 206, 208
"standard," 18-19, 205-206
See also Subcontracting; Franchising

254 / SAFETY OR PROFIT?

Enforceable undertakings, 125-127
Ergonomics, 42
European Union
  European Community Strategy on
    Health and Safety at Work, 211
  EU Framework Directive 89/391, 58,
    210-211
  other EU Directives, 58
  OHS policies and strategies, 209-212
  Working Time Directive, 150-153
Evidence based policy, 6-7, 65, 137-156
  passim, 212-214
Experience rating, 91, 169, 172
External responsibility system, 7, 76, 168

Factories and Shops Act (Australia), 115
Factory Regulation Act (UK), 115
Flexibility, 16, 17, 18, 20, 26, 31, 206
Franchising, 23, 54

Globalization, 1, 16-17, 23, 27, 28, 53, 78,
    81, 84, 174, 182
Gender, 6, 8, 15, 21, 24, 33-48, 206
  regulatory challenges, 40-44

Ham Report, 76, 81, 159, 160, 162, 174,
    178, 180
Hampton Review, 102, 107, 109, 110
Health and Safety at Work (HSW) Act, 3,
    5, 96, 128, 159, 190-191
Health and safety committees (*see* Worker
    representation and consultation)
"Health and safety madness," 7
Hours of work, 151
  irregular, 22
  night work, 44, 147-151, 154
  shift work, 18
  HRM policies, 67-68
HSE/HSC, 98-111, 190

International Agency for Research on
    Cancer (IARC), 142, 146, 147,
    149-150
ILO, 11, 34, 37, 151, 214

Industries/Occupations
  agriculture, 19, 21, 36
  aviation, 150
  clerical, 36
  clothing, 37
  construction, 19, 21, 36, 57, 58, 159
  dockwork, 19, 21
  domestic, 33, 38
  fishermen, 21
  food, 23
  forestry, 36
  healthcare, 36, 37, 38
  hotel, 24, 59
  manufacturing, 36, 45, 59
  merchant seafarers, 19, 21
  mining, 19, 34, 58
  nursing, 61, 150
  oil and gas, 181-203
  restaurant, 61
  sales/retail, 36, 37
  services, 36
  social work, 36
  sweated trades, 19-22
  teaching, 36, 37, 61
  transport, 19, 23
Injuries/ill health, work-related, 42, 51,
    137-156
  extent of, 1-2, 41, 100, 140, 141-142
Inspectors/inspections, 40, 87, 102, 103,
    105, 107, 110-111, 114, 115, 150, 160
  inspectors, cut backs in, 26, 52, 66,
    101-107, 206-208
Intensification of labour, 6, 17, 24, 26, 28,
    29, 39, 47
Internal responsibility system, 157, 159,
    166, 167, 168, 171, 176, 215

Job insecurity, 18, 21, 25, 28, 29, 30, 47,
    84, 164, 167
  See also Employment contracts

Labour market, 4, 6, 8-9, 12, 17-18, 21,
    23, 25-27, 51, 53, 59, 63, 93-96
Labour Standards Act (Canada), 43
Lean production, 17, 70
Löfstedt Report, 111, 208

INDEX / 255

Maquiladoras, 34
Mass hysteria, 39
Migrant labor, 18, 19, 24, 38, 54, 60, 64,
    89, 160, 161, 170, 215
Minerals Management Service (US), 187,
    192, 193-197, 199-200
Models of work effects on health, 27-31

National Institute for Working Life
    (Sweden), 54, 66
Neo-liberalism, 2, 5-10, 13, 16-28, 31,
    42, 44, 49, 72-74, 77-80, 85, 86,
    93, 95, 101, 109, 124, 138, 142,
    143, 147, 154, 204, 207, 211, 212,
    215
    EU and, 209-212
New governance theory, 9-10, 49, 71-95,
    209, 215
New Labour (UK), 77, 97-111, 207
NHS, 143, 145, 146

Occidental Petroleum, 184, 185, 186,
    191
Occupational Health and Safety Act
    (Ontario), 74, 173
Occupational Health and Safety
    Administration (US), 79, 83
Occupational Health and Safety Act
    (Québec), 35
Occupational Health and Safety Act
    (NSW, Aus), 132
Occupational Health and Safety Act 1985
    (Victoria, Aus.), 117, 122
Occupational Health and Safety Act 2004
    (Victoria, Aus.), 131
Occupational health (and safety)
    services/specialists, 145, 151-152,
    165, 167, 175, 179, 208

Participation (*see* Worker representation
    and consultation)
Penalties (*see* Regulation)
Piper Alpha, 135, 181, 183, 184, 185
Political economy of speed, 13, 183,
    209

Precarious work, 17-31, 38-39, 139, 160,
    206
    categories of, 24
    externalities of, 25-26
    historical contexts, 18-21
Precautionary approach to public health,
    134, 135, 141, 149, 153, 212, 213
Preventative services (*see* Occupational
    health (and safety) services/
    specialists)
Private litigation, 85
Privatization, 17, 23, 39, 59, 204, 206
Psychosocial hazards, 6, 23, 26, 27, 40,
    41, 42, 47, 51, 55, 60, 64-65, 69, 174
Publicity, as corporate sanction, 125, 129

Reasonably practicable, as qualification to
    liability, 116, 121
Refusal of work, 90, 91
Regulation
    "burden" of, 2, 7, 42, 100-102
    deregulation, 9, 13, 79, 83, 101, 104,
        204, 209
    fines (penalties), 11, 86, 87, 88, 89, 90,
        93, 112, 113, 116, 117, 119, 120, 121,
        125, 126-129, 138, 158, 168, 169,
        170, 176, 177, 201, 202, 208
    prosecution, 4, 11, 49, 75, 76, 79, 87,
        89, 90, 95, 103, 113-133, 150, 158,
        168, 169, 176, 177, 179, 207, 209
    regulated self-regulation, 78, 79, 85,
        94
    risk based, 109, 123, 124
    "self-regulation," 3, 5, 9, 10, 49, 71, 72,
        75, 76, 78, 83, 85, 86, 93, 94, 95, 97,
        102, 107, 110, 135, 158-159, 164,
        197, 208
Regulatory capture, 134, 185, 190-194,
    209
Regulatory exceptionalism, zone of,
    182-184, 190, 194, 203
Regulatory impact assessment, 8, 100,
    208, 213
Regulatory Reform Act (UK), 100
Regulatory surrender, 103
Responsive enforcement, 123, 124, 126

256 / SAFETY OR PROFIT?

Risk assessment, 57, 63, 146, 150, 152, 196, 199, 210
Robens report, 3, 4-8, 9, 10, 45, 47, 49, 71-72, 75, 77, 78, 80, 93, 94, 97-99, 103-104, 110, 113-114, 121-122, 135, 157-159, 163, 180, 189, 207, 208, 209

Soft law, 42, 43, 211
Stop work orders, 86, 87, 116, 169, 170
Strict liability, 115, 208
Structuration of failure, 181, 203
Structures of vulnerability, 204, 206
Subcontracting, 17, 18, 20, 21, 24, 25, 29, 31, 54, 57, 161, 206
Supply chains, 17, 23, 54, 57, 84, 160, 162, 177

Temporary employment agencies, 38, 43
Trade unions
  contribution to OHS, 7
  and occupational cancer, 12, 137, 138, 140, 143, 146, 152, 155
  reduced strength, 6, 205, 212
  and tripartism, 5, 104-107
Toronto Workers' Health and Safety Legal Clinic, 163, 167
Training, 65-66, 69, 161
Triangle Shirtwaist factory fire, 33
Tripartite organisation, 5, 104-107, 109, 110, 146, 151, 215
TUC, 110, 151, 153

Victimization/harassment, 40, 42, 43, 55, 56, 57, 64, 169, 202-203
Violence at work, 40, 42, 43, 55, 64, 138
Vulnerable workers, 18, 38, 85, 94, 107, 139, 145, 146, 151, 155, 160, 162, 164, 167, 169, 170, 173, 174, 176-177, 215

Whistle blowing, 84, 143
WHO, 23, 141, 155
Women, 28, 33-48, 148, 206
  women's work, 35-40
Work (dis)organization, 29, 30
Work Environment Act (Sweden), 52, 57
Work Environment Authority (Sweden), 52, 54, 62, 63, 66
Work Health and Safety Bill (Model) (Australia), 122, 126, 129, 130, 131, 132
Work/Life balance, 21, 23, 30, 44
WHSA (Workplace Health and Safety Agency), 174
Workers' representation and consultation (Sweden), 50-71
  numbers/coverage of health and safety representatives (Sweden), 55
  influence of safety representatives (Sweden), 62-64, 215
  rights of safety representatives (Sweden), 56-58
  roving/regional safety representatives (Sweden), 59, 61, 94
  rights of safety representatives (other than in Sweden), 7, 9, 41, 74, 169
  and small workplaces, 76
  provisional improvements notices (PINS), 132-133
  training of safety representatives, 173, 176
  union and non-union safety representatives, 5
  barriers to actions by safety representatives, 173
  joint health and safety committees, 5-6, 40, 41, 74, 84, 159, 162, 166, 173, 215
"Worker"/"employee" distinction, 26